SpringerBriefs in Applied Sciences and Technology

SpringerBriefs present concise summaries of cutting-edge research and practical applications across a wide spectrum of fields. Featuring compact volumes of 50 to 125 pages, the series covers a range of content from professional to academic.

Typical publications can be:

- A timely report of state-of-the art methods
- An introduction to or a manual for the application of mathematical or computer techniques
- A bridge between new research results, as published in journal articles
- A snapshot of a hot or emerging topic
- An in-depth case study
- A presentation of core concepts that students must understand in order to make independent contributions

SpringerBriefs are characterized by fast, global electronic dissemination, standard publishing contracts, standardized manuscript preparation and formatting guidelines, and expedited production schedules.

On the one hand, **SpringerBriefs in Applied Sciences and Technology** are devoted to the publication of fundamentals and applications within the different classical engineering disciplines as well as in interdisciplinary fields that recently emerged between these areas. On the other hand, as the boundary separating fundamental research and applied technology is more and more dissolving, this series is particularly open to trans-disciplinary topics between fundamental science and engineering.

Indexed by EI-Compendex, SCOPUS and Springerlink.

More information about this series at http://www.springer.com/series/8884

Majeed Mohamed · Vikalp Dongare

Aircraft Aerodynamic Parameter Estimation from Flight Data Using Neural Partial Differentiation

 Springer

Majeed Mohamed
Flight Mechanics and Control Division
CSIR-National Aerospace Laboratories
Bengaluru, Karnataka, India

Vikalp Dongare
WNS Global Services
Gurugram, India

ISSN 2191-530X   ISSN 2191-5318 (electronic)
SpringerBriefs in Applied Sciences and Technology
ISBN 978-981-16-0103-3   ISBN 978-981-16-0104-0 (eBook)
https://doi.org/10.1007/978-981-16-0104-0

This Springer imprint is published by the registered company Springer Nature Singapore Pte Ltd.
The registered company address is: 152 Beach Road, #21-01/04 Gateway East, Singapore 189721, Singapore

Preface

Aircraft system identification is a powerful technique for building accurate models of aircraft system from noisy data. It consists of three basic steps which are interrelated: (1) the design of an experiment or flight testing; (2) the construction of model from physical laws or black box model; and (3) the estimation of the model parameter from the measurements. Consequently, aircraft modeling requires a relevant parameter estimation tool. This monograph presents neural partial differentiation as an estimation algorithm for extracting aerodynamic derivatives from flight data. A neural modeling of aircraft system has also been discussed. The neural partial differentiation approach can estimate parameters with their statistical information from the noisy data. Moreover, this method avoids the need for the prior information of the aircraft model parameters.

The objective of this monograph is to extend the use of neural partial differentiation method to the multi-input and multi-output (MIMO) aircraft system for the online estimation of aircraft parameters from established neural model. This approach will be relevant for the design of adaptive flight control system. The estimation of aerodynamic derivatives of rigid and flexible aircrafts is treated separately. The longitudinal and lateral-directional derivatives of aircraft are estimated from flight data, and they are discussed in Chaps. 3 and 4, respectively. Besides the aerodynamic derivatives, mode shape parameters of flexible aircraft are also identified as part of identification for the state space aircraft model in Chap. 5. The contents of this book are intended for readers who want to pursue the postgraduate and doctoral degree in science and engineering. Also, it should be useful to practicing scientists, engineers, and teachers in the field of aerospace engineering.

The completion of this monograph could not have been accomplished without the support of our family members. Our sincere thanks go to anonymous reviewers who made valuable suggestions and gave review of the research work to bring this level of excellence.

Bengaluru, India
November 2020

Majeed Mohamed
Vikalp Dongare

Contents

About the Authors

Majeed Mohamed did his degree in Instrumentation and Control Engineering and M.Tech. in Control Systems from IIT Delhi in 2002 and completed his Ph.D. in Flight Dynamics and Control from IIT Delhi in 2012. He is presently working as Principal Scientist in Flight Mechanics and Control Division at National Aerospace Laboratories (NAL) Bangalore and has worked with CSIR for 21 years. Dr. Majeed is the recipient of a research fellowship award from Nanyang Technological University (NTU), Singapore, for his postdoctoral work in 2016 at the Air Traffic Management Research Institute (ATMRI), Singapore. He has authored the book 'Aircraft System Identification and Control' in 2013. Dr. Majeed has published 17 international journal papers and 20 international papers in IFAC and IEEE conferences. He has guided over ten M.Tech. students of flight dynamics and control. He is also Associate Professor of the Academy of Scientific and Innovative Research (AcSIR), New Delhi, India.

Vikalp Dongare did his degree in Avionics System and Engineering in 2012 from Aeronautical Society of India, New Delhi, and M.Tech. in Aeronautical Engineering from Visvesvaraya Technological University, Bangalore, in 2015. He has completed an internship of one year from CSIR-National Aerospace Laboratories Bangalore and has several journal and conference proceedings publications. He is a life member of the Aeronautical Society of India. Vikalp is presently working as a Data Scientist in a multinational corporation to build advanced analytical models for aviation and healthcare businesses. He has experience in making big data analytics and machine learning models.

Acronyms

AoA	Angle of Attack
AoS	Angle of side-slip
BC	Boundary Condition
EEM	Equation Error Method
EKF	Extended Kalman Filter
FEM	Filter Error Method
IC	Initial Condition
MIMO	Multi-Input and Multi-Output
MISO	Multi-Input and Single-Output
NPD	Neural Partial Differentiation
NN	Neural Network
ODE	Ordinary Differential Equations
OEM	Output Error Method
PDE	Partial Differential Equation
RSD	Relative Standard Deviation
SD	Standard Deviation
SI	System Identification
UKF	Unscented Kalman Filter

Chapter 1
Aircraft System Identification

1.1 Introduction

In a fairly complex system like aircraft, modeling and parameter estimation play a crucial role in determining its stability and control characteristics. Applications of the parameter estimation method to estimate such parameters from flight data in the linear flight regime have been highly successful in the past (Jategaonkar 2006; Klein and Moreli 2006). The neural modeling represents an appealing alternative for aircraft system modeling and control mainly because neural network can learn nonlinear input/output mappings from flight data. A neural network brings important benefits of suppressing theoretical difficulties that appear when applying classical techniques on aircraft systems including nonlinearities in their structure. As a result, neural network can describe or control accurately aircraft nonlinear systems with few a priori theoretical knowledge, and it is able to find many successful applications of neural network in the Aerospace industry. The neural-network-based adaptive flight controller for uncertain, nonlinear dynamical systems eliminates the need for offline gain tuning and scheduling methods (Pashilkar et al. 2006; Pesonen et al. 2004). A fault-tolerance, Neural-network-based algorithm was also successfully applied for the flush data sensing system (Rohloff et al. 1999). Many researchers were showed that aircraft neural model has the potential to accommodate changes in aircraft dynamics due to system uncertainties (Ghosh et al. 1998; Hornik et al. 1989).

Parameter estimation from flight data, as applied to aircraft in the linear flight regime, is currently being used on a routine basis with the assumption that the rigid body model is valid (Eugene 2006). Elastic degrees of freedom are, therefore, absent from the aircraft derivative model used in the estimation algorithm. The aircraft with a high degree of flexibility may yield to system dynamics that contain too many parameters, which are required to be estimated. The estimation of rigid body and elastic body derivatives is also demonstrated in Colin et al. (2008), Majeed (2014b), Majeed and Jatinder (2013), Raisinghani and Ghosh (2000), Samal et al. (2009) for a valid model of flexible aircraft in a wide range of frequency. However, the recently

© The Author(s), under exclusive license to Springer Nature Singapore Pte Ltd. 2021   1
M. Mohamed and V. Dongare, *Aircraft Aerodynamic Parameter Estimation from Flight Data Using Neural Partial Differentiation*,
SpringerBriefs in Applied Sciences and Technology,
https://doi.org/10.1007/978-981-16-0104-0_1

introduced neural partial differential method is able to give theoretical insight into statistical information of relative standard deviation (RSTD) of estimates from noisy data (Das et al. 2010; Sinha et al. 2013). In the presence of change in process noise, an adaptive unscented Kalman filter is used to estimate aerodynamic parameters accurately from flight data (Majeed and Kar 2013), but this algorithm requires high computational power and initial values of the estimates.

Since the aircraft is a complex, highly coupled, and nonlinear system, it needs a higher order modeling method to completely describe the system. The commonly used methods to estimate aircraft parameters are output error method (OEM) and filtering methods. They need a priori knowledge of dynamic model and initial values of aerodynamic parameters to estimate aircraft stability and control parameter (Eugene 2006; Field et al. 2004; Grauer 2015; Majeed and Kar 2013). The initial values of parameters for rigid body aircraft are mostly available from the wind tunnel database. But, a lot of aeroelastic derivatives are involved in the flexible aircraft model and therefore, initial values of certain derivatives are not known. Conversely, the use of a scaled version of aircraft in the wind tunnel may introduce the errors in the prediction of aerodynamic derivatives and those derivatives will be using as initial value of parameters for estimation algorithm. This requires the application of an alternative method that can provide accurate estimates of aircraft parameters without their initial values (Jategaonkar 2006; Colin et al. 2008).

The equation error method (EEM) is an alternative method, but can only be used for deterministic system as opposed to the stochastic approach of filtering method. This is considered as the simplest method for the aircraft parameter estimation, which was successfully applied to flight data of F-18 High alpha research vehicle (Morelli 2000), and to a gliding flight vehicle (Umit et al. 2009). EEM needs only the model structure of the aircraft dynamics and does not require initial values of parameters. However, in the presence of noise, the least squares estimates of EEM are asymptotically biased, inconsistent, and inefficient (Eugene 2006). Therefore, neural networks have become a preferred alternative for aerodynamic parameter estimation as they do not require a priori information about the model structure and the parameters of aircraft system dynamics (Calia et al. 2008; Kumar 2012; Peyada and Ghosh 2009; Raisinghani and Ghosh 2000, 2001; Raisinghani et al. 1998; Singh and Ghosh 2007; Samal et al. 2008; Singh and Ghosh 2013). A class of neural network called the feed forward neural networks (FNNs) has been used to model aircraft dynamics wherein aircraft motion variables and control inputs are mapped to predict the total aerodynamic coefficients (Hess 1993; Linse and Stengel 1993; Shi et al. 2007). The capability of FNN for aerodynamic modeling of a flexible aircraft and the applicability of the delta method and the lambda gamma learning rule for extracting parameters from a neural model are demonstrated (Raisinghani and Ghosh 2000; Samal et al. 2009).

The mathematical model of the dynamical system either has a linear or nonlinear structure. Delta and zero method of neural networks can be used to extract aircraft aerodynamic derivatives from flight data irrespective of model structure (Ghosh et al. 1998; Raisinghani et al. 1998). These methods provide the estimate of aircraft parameters, but the statistics of estimates are not inferred directly. Whereas, NPD method (Sinha et al. 2013) is used to estimate flight stability and control parameters with their relative standard deviation. This method has been originated from the fact that the solution of the ordinary differential equation and the partial differential equation can be obtained by neural networks (Fojdl and Brause 2008; Lagaris et al. 1998). Moreover, NPD method can also extract parameters of dynamical systems, which are appearing as nonlinear to the states of the system. This thesis extends the use of NPD method for multi-input and multi-output (MIMO) aircraft system, previously it was employed only for multi-input and single-output system (MISO) (Das et al. 2010).

1.2 Key Aspects of Aircraft System Identification

Aircraft system identification is the determination of an accurate and validated mathematical model of flight vehicles. This is a necessary step in flight vehicle development; because the derived mathematical model is required for (i) understanding the cause-effect relationship, (ii) investigating aircraft performance and characteristics, (iii) verifying aerodynamic database, (iv) updating flight control law design, (v) supporting for flight envelope expansion, (vi) reconstructing the flight path trajectory, including wind estimation and incidence analysis, and (vii) performing fault-

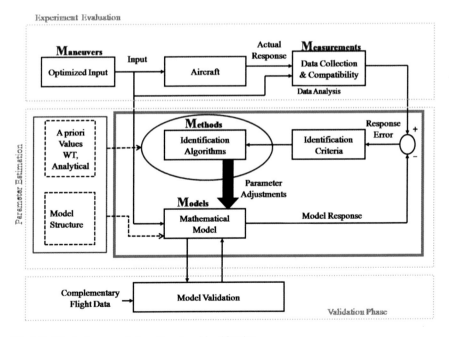

Fig. 1.1 Block diagram of aircraft system identification

diagnosis and adaptive control. The procedure for aircraft system identification is presented in Fig. 1.1.

The aircraft dynamics are modeled by a set of differential equations representing aircraft equation of motion. The external forces and moments acting on the aircraft are described in terms of aircraft stability and control derivatives, which are treated as unknown parameters of the postulated mathematical model. Using specifically designed control inputs to aircraft, actual aircraft response and model response are compared. The difference between actual response and model response gives the parameter estimation algorithms that are applied to minimize this response error by iteratively adjusting the model parameters.

Thus, the key elements for aircraft system identification are manoeuvres, measurements, methods, and models. The aspects of these elements are referred to as Quad-M requirements and aircraft system identification and they have to be treated carefully (Hess 1993). This is because of identification results depend on (1) optimal input (manoeuvres) to excite different modes of the aircraft system, (2) accuracy of aircraft air data system, (3) selection of the most suitable parameter estimation methods, and (4) structure of the mathematical model to represent the best relationship between the model complexity and measurement available. Model verification is the final step in the model building process and it should be carried out no matter how sophisticated estimation algorithm is applied. Model predictive capability is the most widely used procedure for verification of the flight-estimated models. For verifica-

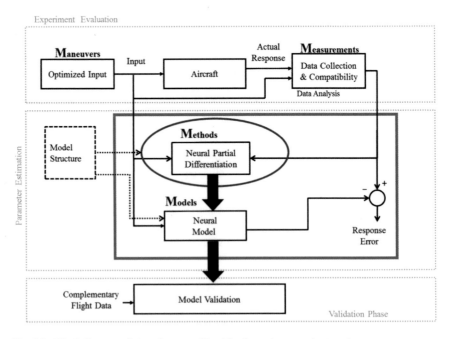

Fig. 1.2 Block diagram of aircraft system identification using neural network

tion, the model parameters are fixed to the estimated values and the model is driven by inputs that are different from those used in estimation. The model parameters are compared with the flight measurements to check upon the predictive capability of the estimated model. If we choose the NPD as parameter estimation method, the initial value of the parameter is not required to estimate the system parameter, moreover, the NPD approach has the potential to estimate the parameter from noisy data with their noise statistics. Therefore, it gives relaxation of the accurate measurement of air data sensors. The block diagram for application of NPD of aircraft parameter estimation is shown in Fig. 1.2.

1.3   History

Aircraft parameter estimation is probably the most outstanding and illustrated example of the identification of the aircraft system from flight data. The highly successful application of aircraft system identification has been possible due to (1) the better measurement techniques and data processing capabilities provided by the digital computers, (2) the ingenuity of engineers in advantageously using the developments in other fields such as estimation and control theory, and (3) the fairly well understood basic physical principles leading to adequate aerodynamic modeling, and design of appropriate flight tests (Jategaonkar 2006; Eugene et al. 2013; Maine and Iliff 1986). The four important aspects, namely excitation input, measurements, parameter estimation methods, and mathematical model are needed to be treated carefully in the area of system identification. Several authors made substantial contributions in the field of system identification applied to aircraft in the 1960s and 1970s. These contributions made the area of development and application of various estimation techniques (Colin et al. 2008; Joao et al. 2005; Klein and Moreli 2006; Klein and Batterson 1985). In these contribution, "Mulder" addressed the method for experiment design, measured data compatibility, and parameter estimation (Joao et al. 2005; Mulder 1986).

The first analytical technique used to estimate aircraft dynamic model parameters from flight data was the equation error method (EEM). This method represents a broad class of methods that are applicable to linear time-invariant dynamic systems. It is a regression approach that requires direct measurement of all state variables. It constitutes the dynamic equations, linear in terms of unknown parameters (Eugene 2006; Umit et al. 2009). Practical aspects of the EEM are used for the purpose of aircraft parameter estimation, which are available in Eugene (2006), Umit et al. (2009). Even though EEM is a single short solution for parameter estimation method, this approach cannot handle the presence of noise in the flight data (Jategaonkar and Jategaonkar 1989; Kutluay and Mahmutyazicioglu 2009; Li et al. 2013). Therefore, this method is advisable only if high-quality sensors are used for measuring the aircraft system responses. Whereas the output error method (OEM) is widely accepted for aircraft parameter estimation in the presence of measurement noise. OEM approach and its most employed version in various aspects of the Maximum Likelihood (ML) method are well documented by "Mehara and stepner" (Mehra 1970; Stepner and Mehra 1973). On the other hand, the generality and usefulness of the equation error approach are highlighted by the recent analytical developments for the special solution of aircraft dynamic modeling (Bucharles et al. 2012; Kalman 1960).

Normally, a post-flight batch technique was applied to solve the data-compatibility problem using output error parameter estimation approach (Iliff and Maine 1983; Jategaonkar et al. 2004). Short segments of data were sequentially processed to enable real-time estimation, and variations on the algorithm are used to expedite convergence from arbitrary starting values of the unknown model parameters. The method was applied to flight test data to correct the data for systematic instrumentation errors (Grauer 2015; Joao et al. 2005). OEM method is considered as a simple and iterative solution to the parameter estimation problem in the presence of measurement noise. But this method cannot handle process noise in the dynamical system and also required the initial values of parameters. Therefore, under these negative circumstances, this approach sometimes may not be ensuring the guarantee of convergence within the error bound defined by an user (Joao et al. 2005; Majeed 2014b; Majeed and Kar 2011; Li et al. 2013).

The filter error method (FEM) method has a significant advantage in providing parameter estimates in the presence of the process noise. It is based on the Kalman filter for providing the state estimations of the identified system. This FEM approach is some extension of the output error method. It can take care of both measurement and process noise of the system. It may be effectively applied to the nonlinear system and unstable aircraft (Chowdhary and Jategaonkar 2010). Although it's highly dependent on stochastic process noises model, it required guess values of parameters. There are several studies available that utilize the filter error method for estimating aircraft parameters from the flight data in the presence of turbulence (Y 1977). But this method is highly complex for aircraft system if the computational burden is considered.

The Extended Kalman Filter (EKF) is the most popular nonlinear filtering technique in the aerospace industry, which employs instantaneous linearization at each time step to approximate the nonlinearities. However, the EKF can be hard to tune and implement when dealing with significant nonlinearities and exhibits divergence in extreme cases. Despite its theoretical shortcomings, however, the EKF has been employed successfully in various aircraft aerodynamic parameter estimation problems. The problems associated with the EKF are attributed to the approximation introduced by the linearization. The EKF approximates the mean and covariance using a first-order approximation of the system dynamics. The Unscented Kalman Filter (UKF) overcomes these theoretical deficiencies by using a set of carefully selected sample points (also known as Sigma Points) to approximate the probability distribution of the random variable (Majeed and Kar 2013; Majeed 2014a).

Since an aircraft system is complex and nonlinear, it needs a higher degree of modeling. The neural network can be used as a black box modeling and establishes a neural model of the dynamical system. The online parameter estimation can be carried out from such a neural model. This approach does not require a priori model of the dynamic system and can handle the higher computational cost for a complex mathematics model. Unlike delta-zero and Neural-Gauss-Newton (NGN) method, NPD method can estimate the stability and control derivative with noise statics in the data. However, the estimation of the noise statics is a tedious job in delta-zero method (Sinha et al. 2013). Whereas the introduced NPD method can estimate aircraft stability and control derivative from noisy data with their noise statistics.

Neural-network-based delta and zero methods (Raisinghani and Ghosh 2000; Raisinghani et al. 1998) showed potential to estimate parameters of dynamics system. Both these methods apply finite differencing to obtain the stability and control derivatives in the post-training phase. As an extension to this work, the modified delta method (Singh and Ghosh 2007) was developed to yield the parameter estimates with a small standard deviation. However, all these methods require additional data processing in the post-training phase in order to extract the stability and control derivatives by finite difference approximation. A combination of the neural network and the Gauss-Newton method (Kumar 2012; Taylor et al. 1969) can also be applied to estimate the stability and control derivatives. In this method, the neural network is used to propagate the state of the aircraft, which is followed by application of the Gauss-Newton method to estimate the stability and control derivatives by minimizing a suitable cost function. However, all these methods lack a precise theoretical basis and do not give any insight into the nature of the results.

This differs from the earlier neural-network-based methods, in that the sigmoid function was replaced by the radial basis function. The methodology is exactly the same as for the other neural methods mentioned earlier, and therefore requires data processing in the post-training phase. However, this method requires a large number of hidden neurons and is slower than the sigmoid nonlinearity-based neural network (Hess 1993). To avoid the need of post-training data processing and to improve the accuracy of estimation, a neural partial differentiation method (Raisinghani and Ghosh 2000) was reported for the estimation of the lateral/directional stability and control derivatives from the actual flight data of an aerodynamically stable aircraft. The results obtained using the neural partial differentiation method showed superior performance compared to those obtained using the zero, delta, and least squares methods on actual flight data. However, the work presented did not show theoretical developments on the accuracy and nature of the estimates obtained using the neural partial differentiation approach. Moreover, the effect of the measurement noise in the input and output variables was not theoretically developed, and it was not investigated through simulation. Unlike the other neural methods reported in literature, the neural partial differentiation approach can give theoretical insight into the statistical information, such as the relative standard deviation (RSTD) that is equivalent to the Cramer-Rao bound in output error method. Moreover, the neural partial differentiation approach is superior in terms of computational efficiency and uncertainty in the estimates compared to the other neural methods.

1.4 Outline

The organization of this book is as follows: Chap. 2 describes the neural modeling and parameter estimation of the dynamic systems. This chapter describes the neural partial differentiation approach for the purpose of parameter estimation and analysis of the estimates. The main goal of the investigation is to obtain online estimates of system parameters. The rest of the chapter describes aerodynamic parameter estimation and analysis of those estimates in detail.

The online estimation of aerodynamic parameters of an aircraft longitudinal dynamics is presented in Chap. 3. The investigation is initially made on simulated data with additive noise, and it is shown that all estimated parameters are accurate. A comparative study has been done between NPD and OEM for the estimation of flight stability and control parameter from flight data. It is found that NPD successfully attains better estimates.

NPD method is further applied to estimate lateral-directional derivatives of an aircraft, and it is presented in Chap. 4. Primarily, NPD method is applied for the simulated data with additive noise and it is shown that all the parameters estimated are accurate. A comparative study has been done between NPD and OEM for the estimation of flight stability and control parameter from flight data. It was found that NPD attains better estimates without knowing their initial values.

References

A. Bucharles, C. Cumer, G. Hardier, B. Jacquier, A. Janot, T.L. Moing, P. Vacher, An overview of relevant issues for aircraft model identification. J. Aerosp. lab. **4**, 1–21 (2012)

A. Calia, E. Denti, R. Galatolo, F. Schettini, Air data computation using neural networks. J. Aircraf. **45**(6), 2078–2083 (2008)

G. Chowdhary, R. Jategaonkar, Aerodynamic parameter estimation from flight data applying extended and unscented kalman filter. Aerosp. Sci. Technol. 106–117 (2010). https://doi.org/10.1016/j.ast.2009.10.003

T. Colin, I. Christina, T. Mark, F. Edmund, N. Randall, R. Heather, *System Identification of Large Flexible Transport Aircraft*. In Aiaa atmospheric flight mechanics conference and exhibit (American Institute of Aeronautics and Astronautics, 2008). https://doi.org/10.2514/6.2008-6894

S. Das, R.A. Kuttieri, M. Sinha, R. Jategaonkar, Neural partial differential method for extracting aerodynamic derivatives from flight data. J. Guid. Control Dyn. **33**(2), 376–384 (2010). https://doi.org/10.2514/1.46053

A.M. Eugene, C. Kevin, A.H. Melissa, Global aerodynamic modeling for stall/upset recovery training using efficient piloted flight test techniques, in *Aiaa Modeling and Simulation Technologies (mst) Conference* (American Institute of Aeronautics and Astronautics, 2013). https://doi.org/10.2514/6.2013-4976

M. Eugene, Practical aspects of the equation-error method for aircraft parameter estimation, in *Aiaa Atmospheric Flight Mechanics Conference and Exhibit* (American Institute of Aeronautics and Astronautics, 2006). https://doi.org/10.2514/6.2006-6144

E.J. Field, K.F. Rossitto, J. Hodgkinson, Flying qualities applications of frequency responses identified from flight data. J. Aircr. **41**(1), 711–720 (2004)

J. Fojdl, R.W. Brause, The performance of approximating ordinary differential equations by neural nets, in *20th IEEE International Conference on Tools with Artificial Intelligence, 2008. ICTAI '08*, vol. 2 (2008), pp. 457–464. https://doi.org/10.1109/ictai.2008.44

A.K. Ghosh, S.C. Raisinghani, S. Khubchandani, Estimation of aircraft lateral-directional parameters using neural networks. J. Aircr. **35**, 876–881 (1998)

J.A. Grauer, Real-time data-compatibility analysis using output-error parameter estimation. J. Aircr. **52**(3), 940–947 (2015). https://doi.org/10.2514/1.C033182

R.A. Hess, On the use of backpropagation with feed forward neural networks for the aerodynamic estimation problem. AIAA 93–3639 (1993)

K. Hornik, M. Stinchcombe, H. White, Multi layer feed forward neural networks are universal approximators. Neural Netw. **2**, 359–366 (1989)

K.W. Iliff, R.E. Maine, Uses of parameter estimation in flight test [Journal Article]. Journal of Aircraft **20**(12), 1043–1049 (1983). https://doi.org/10.2514/3.48210

R. Jategaonkar, Flight vehicle system identification: a time domain methodology (American Institute of Aeronautics, 2006). Astronautics

R. Jategaonkar, D. Fischenberg, W.V. Gruenhagen, Aerodynamic modeling and system identification from flight data-recent applications at dlr. J. Aircr. **41**(1) (2004)

R. Jategaonkar, E. Plaetschke, Algorithm for aircraft parameter estimation accounting for process and measurement noise. J. Aircr. **26**(4), 360–372 (1989)

Joao, O., Chu, Q. P., Mulder, J. A., Balini, H. M. N. K., & Vos, W. G. M. (2005). Output error method and two step method for aerodynamic model identification [Book Section]. *In Aiaa guidance, navigation, and control conference and exhibit*. American Institute of Aeronautics and Astronautics. https://doi.org/10.2514/6.2005-6440

K, Y. (1977). Identification of aircraft stability and control derivatives in the presence of turbulence. *In Guidance and control conference*. American Institute of Aeronautics and Astronautics. https://doi.org/10.2514/6.1977-1134

R.E. Kalman, A new approach to linear filtering and prediction problems [Journal Article]. Journal of Fluids Engineering **82**(1), 35–45 (1960). https://doi.org/10.1115/1.3662552

Vladislav Klein, E.A. Moreli, *Aircraft system identification theory and practice [Book]* (AIAA, Education series, Reston, VA, 2006)

V. Klein, J. Batterson, Aerodynamic parameters estimated from flight and wind tunnel data [Journal Article]. Journal of Aircraft **23**(4), 306–312 (1985)

Kumar, R. (2012). Lateral parameter estimation using ngn method [Journal Article]. *International Journal of Engineering Inventions*, 1(7), 82-89. 10 1 Aircraft system identification

U. Kutluay, G. Mahmutyazicioglu, An application of equation error method to aerodynamic model identification and parameter estimation of a gliding flight vehicle [Conference Proceedings]. Aiaa atmospheric flight mechanics conference. AIAA 2009–5724 (2009)

I.E. Lagaris, A. Likas, D.I. Fotiadis, Artificial neural networks for solving ordinary and partial differential equations [Journal Article]. IEEE Transactions on Neural Networks **9**(5), 987–100 (1998)

Li, C., Dou, X., & Wu, L. (2013). Research of the aerodynamic parameter estimation for the small unmanned aerial vehicle [Conference Proceedings]. In Control and automation (icca), 2013 10th *ieee international conference on* (p. 1917- 1920). https://doi.org/10.1109/ICCA.2013.6565091

D.J. Linse, R.F. Stengel, Idetification of aerodyanamics coefficient using computational neural network [Journal Article]. Journal of Guidance, Control, and Dynamics **16**(6), 1018–1025 (1993)

R.E. Maine, K.W. Iliff, Identification of dynamic system-application to aircraft, part i. Agard AG-300 (1986)

M. Majeed, *Aircraft System Identification and Control* (LAP LAMBERT Academic Publishing, 2014a)

M. Majeed, Parameter identification of flexible aircraft using frequency domain output error approach, in *International Conference on Advances in Control and Optimization of Dynamical Systems (acods)* (2014b)

M. Majeed, S. Jatinder, Frequency and time domain recursive parameter estimation for a flexible aircraft, in *19th IFAC Symposium on Automatic Control in Aerospace* (2013), pp. 443–448

M. Majeed, I.N. Kar, Identification of aerodynamic derivatives of a flexible aircraft using output error method, in *IEEE International Conference on Mechanical and Aerospace Engineering (cmae-2011)* (2011), pp. 361–365

M. Majeed, I.N. Kar, Aerodynamic parameter estimation using adaptive unscented kalman filter. Aircr. Eng. Aerosp. Technol. **85**(4), 267–279 (2013). https://doi.org/10.1108/AEAT-Mar-2011-0038

R.K. Mehra, Maximum likelihood identification of aircraft parameters, in *The Joint Automatic Control Conference* (1970), pp. 442–444

E. Morelli, Real time parameter estimation in the frequency domain. J. Guid. Control Dyn. **23**(5), 812–818 (2000)

J.A. Mulder, *Design and evaluation of dynamic flight test maneuvers* (Report No. Report LR-497). Delft University of Technology, Department of Aerospace Engineering (1986)

A.A. Pashilkar, N. Sundararajan, P. Saratchandran, Adaptive backstepping neural controller for reconfigurable flight control systems. IEEE Trans. Control Syst. Technol. **14**(3), 553–561 (2006). https://doi.org/10.1109/TCST.2005.863672

U.J. Pesonen, J.E. Steck, K. Rokhsaz, H.S. Bruner, N. Duerksen, Adaptive neural network inverse controller for general aviation safety. J. Guid. Control Dyn. **27**(3), 434–443 (2004). https://doi.org/10.2514/1.1923

N.K. Peyada, A.K. Ghosh, Aircraft parameter estimation using neural network based algorithm. AIAA (2009), Accessed 10–13 August 2009

S. Raisinghani, A. Ghosh, Frequency-domain estimation of parameters from flight data using neural networks. J. Guid. Control Dyn. **24**(3), 525–530 (2001)

S. Raisinghani, A. Ghosh, S. Khubchandani, Estimation of aircraft lateral-directional parameters using neural networks. J. Aircr. **35**(6), 876–881 (1998)

S.C. Raisinghani, A.K. Ghosh, Parameter estimation of an aeroelastic aircraft using neural networks. Sadhana **25**(2), 181–191 (2000)

S.C. Raisinghani, A.K. Ghosh, P.K. Kalra, Two new techniques for aircraft parameter estimation using neural networks. Aeronaut. J. **102**, 25–29 (1998)

T.J. Rohloff, S.A. Whitmore, I. Catton, Fault-tolerant neural network algorithm for flush air data sensing. J. Aircr. **36**(3), 541–549 (1999). https://doi.org/10.2514/2.2489

M.K. Samal, S. Anavatti, M. Garratt, Neural network based system identification for autonomous flight of an eagle helicopter, in *IFAC* (2008). Accessed 6–11 July 2008

M.K. Samal, A. Singhal, A.K. Ghosh, Estimation of equivalent aerodynamic parameters of an aeroelastic aircraft using neural network. IE(I) Journal-AS **90** (2009)

Y. Shi, W. Qian, W. Yan, J. Li, Adaptive depth control for autonomous underwater vehicles based on feedforward neural networks. Int. J. Comput. Sci. Appl. **4**(3), 107–118 (2007)

S. Singh, A.K. Ghosh, Modified delta method for estimation of parameters from flight data of stable and unstable aircraft, in *2013 IEEE 3rd International Advance Computing Conference (IACC)* (2013), pp. 775–781. https://doi.org/10.1109/IAdCC.2013.6514325

M. Sinha, R.A. Kuttieri, S. Chatterjee, Nonlinear and linear unstable aircraft parameter estimations using neural partial differentiation. J. Guid. Control Dyn. **36**(4), 1162–1176 (2013). https://doi.org/10.2514/1.57029

S. Singh, A. Ghosh, Estimation of lateral-directional parameters using neural networks based modified delta method. Aeronaut. J. **111**(3150), 659–667 (2007)

D. Stepner, R. Mehra, *Maximum likelihood identification and optimal input design for identifying aircraft stability and control derivatives* (Report No. NASA CR-2200) (1973)

L.W. Taylor, K.W. Iliff, B. Powers, A comparison of newton-raphson and other methods for determining stability derivatives from flight data. AIAA Paper 69-315. 12 1 Aircraft system identification (1969)

K. Umit, M. Gokmen, P. Bulent, *An application of equation error method to aerodynamic model identification and parameter estimation of a gliding flight vehicle [Book Section]* (American

Institute of Aeronautics and Astronautics, In Aiaa atmospheric flight mechanics conference, 2009). https://doi.org/10.2514/6.2009-5724

Chapter 2
Neural Modeling and Parameter Estimation

The neural modeling of a dynamic system is presented in this chapter. The former literature reported that ordinary differential equations can be solved by an neural-network-based approach (Lagaris et al. 1998). The same conceptual materials help to solve the set of partial differential equations (Das et al. 2010). These theoretical materials of neural network (NN) are presented in Appendix for the ready reference. The following section describes how we can estimate parameters from the neural output of a dynamical model.

2.1 Neural Modeling

The neural networks are made up of two main components namely neurons or nodes and the connectors. The connectors have their own weights between two nodes. The neural network uses the data set of input and output, to map the function on to the network in the form of weights between the internal nodes as shown in the Fig. 2.1. The schematic structure of a three-layered feed forward neural network (NN) is consisting of two hidden layers with activation function and one output layer with summation function exempted from activation function. The weights indirectly represent the function of a given system for which the neural network is trained. The output of each node is the sum of the product of the total input to the particular node and their respective weights, applied to an activation function. The back-propagation approach is used for training the neural network. The neural networks learn through the input-output pair of the system and give an approximate function in the form of weights. The complexity of the network can be changed with the number of neurons and/or the number of hidden layers, this decision is purely based on the trial and error method (Hess 1993). The input and output vectors of neural network are defined as

© The Author(s), under exclusive license to Springer Nature Singapore Pte Ltd. 2021   13
M. Mohamed and V. Dongare, *Aircraft Aerodynamic Parameter Estimation from Flight Data Using Neural Partial Differentiation*,
SpringerBriefs in Applied Sciences and Technology,
https://doi.org/10.1007/978-981-16-0104-0_2

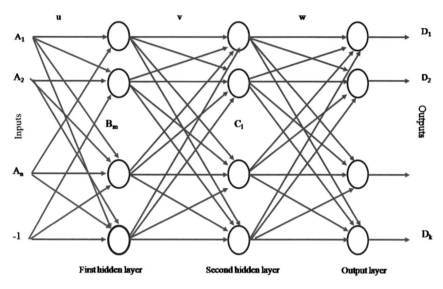

Fig. 2.1 Schematic of neural network

$A \in \mathfrak{R}^{n+1}$ and $D \in \mathfrak{R}^k$, respectively. Similarly, $B \in \mathfrak{R}^{m+1}$ and $C \in \mathfrak{R}^{l+1}$ represent the first and second hidden layer of neural network. Except for the output layer all the layers contain a bias term. Thus, the output of neural network is given by

$$D = W^T C \tag{2.1}$$

where W is the set of weights between the second hidden layer and output layer containing the bias terms.

$$W = \begin{bmatrix} b_{w1} & \cdots & b_{wk} \\ w_{11} & \cdots & w_{1k} \\ \vdots & \ddots & \vdots \\ w_{l1} & \cdots & w_{lk} \end{bmatrix} \tag{2.2}$$

Similarly, we define

$$\begin{cases} C = f(V^T B) \\ B = g(U^T A) \end{cases} \tag{2.3}$$

where f and g are the activation function vectors and are defined as (Dongare and Mohamed 2015)

$f = [-1 \; f(x_1) \; \cdots \; f(x_k)]^T$ where $f = (x)$ is expressed as

$$f(x) = \frac{1 - e^{-\lambda x}}{1 + e^{-\lambda x}} \tag{2.4}$$

And the weight matrices are represented as

$$V = \begin{bmatrix} b_{v1} & \cdots & b_{vm} \\ v_{11} & \cdots & v_{1l} \\ \vdots & \ddots & \vdots \\ v_{m1} & \cdots & v_{ml} \end{bmatrix} \tag{2.5}$$

$$U = \begin{bmatrix} b_{u1} & \cdots & b_{um} \\ u_{11} & \cdots & u_{1m} \\ \vdots & \ddots & \vdots \\ u_{n1} & \cdots & u_{nm} \end{bmatrix} \tag{2.6}$$

Input is defined by the vector $A = [\, a_0 \; a_1 \; \cdots \; a_n \,]$, where a_0 defines bias input to the neural network. The input and output are scaled for neural network using the following equation.

$$D_{i,norm} = D_{i,norm_{min}} + (D_{i,norm_{max}} - D_{i,norm_{min}}) \\ \times \frac{D_i - D_{i,min}}{D_{i,max} - D_{i,min}} \tag{2.7}$$

where $D_{i,norm_{max}}$ and $D_{i,norm_{min}}$ denote the higher and lower limits of scaling range of D_i, respectively. They are set to 0.9 and -0.9, respectively. $D_{i,max}$ and $D_{i,min}$ denote the higher and lower values of D_i.

Using the above notations, output of neural network can be written as

$$D = \{W^T f[V^T g(U^T A)]\} \tag{2.8}$$

2.2 Neural Partial Differentiation

In this approach, the neural network is trained with input and output data so as to map the nonlinear function in the form of weights. The activation functions hold the key for the neural partial difference method. This method does not need extra post processing as the delta and zero method demands (Singh and Ghosh 2013). Moreover, it has the facility to determine the higher order partial derivatives of a nonlinear system (Das et al. 2010; Sinha et al. 2013). The partial differentiation of a system can be computed from the end of the training session of the neural network, and provide aerodynamic derivatives directly as follows:

The input and output of function are mapped after the training session of the neural network. Subsequently, the output variables can be differentiated with respect to input variables. Differentiate Eqs. (2.1) and (2.3), we will have the form of (Mohamed and Dongare 2018)

$$\frac{\partial D}{\partial C} = W^T \tag{2.9}$$

$$\frac{\partial C}{\partial B} = f'(V^T) \tag{2.10}$$

$$\frac{\partial B}{\partial A} = g'(U^T) \tag{2.11}$$

Multiplication of Eqs. (2.9), (2.10), and (2.11) gives

$$\begin{cases} \frac{\partial D}{\partial C} \cdot \frac{\partial C}{\partial B} \cdot \frac{\partial B}{\partial A} = W^T \cdot f' V^T \cdot g' U^T \\ \frac{\partial D}{\partial A} = W^T \cdot f' V^T \cdot g' U^T \end{cases} \tag{2.12}$$

where $f' = diag[0 \ f'_1 \ \cdots \ f'_l]$ and $g' = diag[0 \ g'_1 \ \cdots \ g'_m]$. If the input and output of the neural network are normalized, then

$$\frac{\partial D}{\partial A} = \frac{\partial D}{\partial D_{norm}} \times \frac{\partial D_{norm}}{\partial A_{norm}} \times \frac{\partial A_{norm}}{\partial A} \tag{2.13}$$

The normalized output of neural network can be de-normalized by Eq. (2.13) where

$$\frac{\partial D}{\partial D_{norm}} = \begin{bmatrix} \frac{\partial D_1}{\partial D_{1,norm}} & 0 & \cdots & 0 \\ 0 & \frac{\partial D_2}{\partial D_{2,norm}} & \cdots & 0 \\ \vdots & \vdots & \ddots & \vdots \\ 0 & 0 & \cdots & \frac{\partial D_k}{\partial D_{k,norm}} \end{bmatrix} \tag{2.14}$$

$$\frac{\partial A}{\partial A_{norm}} = \begin{bmatrix} 1 & 0 & \cdots & 0 \\ 0 & \frac{\partial A_{1,norm}}{\partial A_1} & \cdots & 0 \\ \vdots & \vdots & \ddots & \vdots \\ 0 & 0 & \cdots & \frac{\partial A_{n,norm}}{\partial A_n} \end{bmatrix} \tag{2.15}$$

Equations (2.14) and (2.15) can be computed from Eq. (2.7). The terms associated with Eqs. (2.9) to (2.15) be intermediate terms of neural networks while getting it trained. Therefore, there is no extra computation required to compute the aerodynamic derivatives, and they are directly given as:

$$\frac{\partial D}{\partial A} = \begin{bmatrix} \frac{\partial D_1}{\partial A_0} & \cdots & \frac{\partial D_1}{\partial A_n} \\ \vdots & \ddots & \vdots \\ \frac{\partial D_k}{\partial A_0} & \cdots & \frac{\partial D_k}{\partial A_n} \end{bmatrix} \tag{2.16}$$

The standard deviation of estimated parameters in Eq. (2.16) is computed by

$$STD = \sqrt{\frac{\sum\limits_{p=1}^{P}\left[\sum\limits_{m=1}^{M}\sum\limits_{l=1}^{L}(\sum C'_{l_p}v_{lm}w_{kl}D'_{k_p})B'_{m_p}u_{mi} - AVG\right]^2}{P}} \qquad (2.17)$$

where

$$AVG = \frac{\sum\limits_{p=1}^{P}\sum\limits_{m=1}^{M}\sum\limits_{l=1}^{L}(\sum C'_{l_p}v_{lm}w_{kl}D'_{k_p})B'_{m_p}u_{mi}}{P} \qquad (2.18)$$

STD and AVG are standard deviation and average of data points, respectively. The relative standard deviation of estimates is given by

$$RSTD = \frac{STD}{AVG} \times 100\% \qquad (2.19)$$

Similarly, the equations for the second-order partial derivatives, using the notations described earlier, can be written as

$$\frac{\partial}{\partial a}\left[\frac{\partial D}{\partial A}\right] = \frac{\partial}{\partial a}\left[\frac{\partial D}{\partial D_{norm}} \times \frac{\partial D_{norm}}{\partial A_{norm}} \times \frac{\partial A_{norm}}{\partial A}\right] \qquad (2.20)$$

$$\frac{\partial}{\partial a}\left[\frac{\partial D}{\partial A}\right] = \left[\frac{\partial D}{\partial D_{norm}}\right]\frac{\partial}{\partial a}\left[\frac{\partial D}{\partial D_{norm}} \times \frac{\partial D_{norm}}{\partial A_{norm}} \times \frac{\partial A_{norm}}{\partial A}\right]\left[\frac{\partial A_{norm}}{\partial A}\right] \qquad (2.21)$$

$$\frac{\partial}{\partial a}\left[\frac{\partial D}{\partial A}\right] = \left[\frac{\partial D}{\partial D_{norm}}\right]\frac{\partial}{\partial a_{norm}}\left[\frac{\partial D}{\partial D_{norm}} \times \frac{\partial D_{norm}}{\partial A_{norm}} \times \frac{\partial A_{norm}}{\partial A}\right]\left[\frac{\partial A_{norm}}{\partial A}\right]\left[\frac{\partial a_{norm}}{\partial A_{norm}}\right] \qquad (2.22)$$

2.3 Study of the Estimates

To represent the nonlinear model of an aircraft as dynamical system accurately in terms of the aerodynamic force and moment equations, we need to analyze the first and higher order derivatives embedded in the dataset of the model. Let us consider a data set pertaining to a multi-input and single-output (MISO) system whose math model is represented by a polynomial χ:

$$\chi(\alpha, q, \delta_e) = \theta_0 + \theta_1\alpha + \theta_2 q + \theta_3\delta_e + \theta_4\alpha^2 + \theta_5 q^2 + \theta_6\delta_e^2 + \theta_7\alpha q + \theta_8 q\delta_e$$
$$+ \theta_9\delta_e + \theta_{10}\alpha^3 \ldots$$
$$(2.23)$$

where $\theta_i (i = 0, 1, 2, 3, 4 \ldots)$ is a non-dimensional parameter, α is the angle of attack, q is the pitch rate, and δ_e is the elevator deflection angle, which are the inputs to the neural network. The output of neural network is represented by Eq. (2.23). This approach can be useful to get the first and higher order partial derivatives. The first derivative with respect to q can be written as

$$\frac{\partial D}{\partial q} = \frac{\partial \chi}{\partial q} = \theta_2 + 2\theta_5 q + \theta_7 \alpha + \theta_8 \delta_e \dots . \tag{2.24}$$

where D is the output of the neural network and $\frac{\partial \chi}{\partial q}$ is a function of q and the other variables. Since the χ is higher order of polynomial, $\frac{\partial D}{\partial q} \neq \theta_2$ as shown in Eq. (2.24) relatively, it contains the effect of higher order variables. The correct value of the coefficient can be obtained by setting $\alpha = q = \delta_e = 0$ in Eq. (2.24). For the nonlinear model, the second-order partial derivative with respect to q can be written as

$$\frac{\partial^2 D}{\partial q^2} = \frac{\partial^2 \chi}{\partial q^2} = 2\theta_5 + \cdots \tag{2.25}$$

The parameter θ_5 is found by again setting $\alpha = q = \delta_e = 0$ in Eq. (2.25) and can be written as $\theta_5 = \frac{1}{2} \frac{\partial^2 D}{\partial q^2}\Big|_{q=0}$. In general, $\theta_i = \frac{1}{!n} \frac{\partial^n D}{\partial q^n}\Big|_{q=0}$ similarly, the estimates of other equivalent parameters related to the different variables of a higher degree can be possible to achieve.

2.4 Summary

The neural modeling of a dynamic system is established for the application of neural partial differentiation (NPD) method to extract the system parameters from the measured flight data. The particular system dynamics are represented by the Neural modeling. But the selection of the architecture of the network in terms of the number of neurons in the hidden layer, the learning rate, the momentum rate, etc. is not an exact science and one has to resort to trial and error methods to find a suitable network structure for the given data. Estimation of system parameters is explained in a very detailed manner using the NPD approach, and its results are repeatedly used in the upcoming chapters for the purpose of online estimation of aircraft parameters from flight data.

References

S. Das, R.A. Kuttieri, M. Sinha, R. Jategaonkar, Neural partial differential method for extracting aerodynamic derivatives from flight data [Journal Article]. J. Guid. Control Dyn. **33**(2), 376–384 (2010). https://doi.org/10.2514/1.46053

V. Dongare, M. Mohamed, Lateral-directional aerodynamics parameter estimation using neural partial differentiation, in *2015 International Conference on Cognitive Computing and Information Processing (ccip)* (2015), pp. 1–6. https://doi.org/10.1109/CCIP.2015.7100730

R.A. Hess, On the use of backpropagation with feed forward neural networks for the aerodynamic estimation problem [Journal Article]. AIAA, 93–3639 (1993)

I.E. Lagaris, A. Fotiadis, D.I. Likas, Artificial neural networks for solving ordinary and partial differential equations [Journal Article]. IEEE Trans. Neural Netw. **9**(5), 987–100 (1998)

M. Mohamed, V. Dongare, Aircraft neural modeling and parameter estimation using neural partial differentiation. Aircraft Eng. Aerospace Technol. **90**(5), 764–778 (2018). https://doi.org/10.1108/AEAT-02-2016-0021

S. Singh, A.K. Ghosh, Modified delta method for estimation of parameters from flight data of stable and unstable aircraft [Conference Proceedings], in *2013 IEEE 3rd International on Advance Computing Conference (IACC)* (2013), pp. 775–781. https://doi.org/10.1109/IAdCC.2013.6514325

M. Sinha, R.A. Kuttieri, S. Chatterjee, Nonlinear and linear unstable aircraft parameter estimations using neural partial differentiation[Journal Article]. J. Guid. Control Dyn. **36**(4), 1162–1176 (2013). https://doi.org/10.2514/1.57029

Chapter 3
Identification of Aircraft Longitudinal Derivatives

This chapter focuses on the application of neural networks to the problem of aircraft aerodynamic modeling and parameter estimation. The neural modeling and neural partial differentiation (NPD) method, which are presented in Chap. 2, are directly applied here to estimate the longitudinal dynamics of an aircraft system. Since the model of an aircraft system is established through the neural network, extends the use of NPD to multi-input and multi-output(MIMO) system ensures online estimation of each aircraft aerodynamics derivatives. As a result, the application of NPD to estimate aircraft aerodynamic parameters in real-time will apply to the flight control system wherein the parameters of aircraft change in a rapid manner.

3.1 Longitudinal Dynamical Model

The longitudinal dynamics of an aircraft can be represented by a set of differential equation in terms of motion-related variables, say angle of attack α and pitch rate q. The time responses of these variables are obtained by providing an elevator input of δ_e. By using the Taylor series expansion, coefficients C_m and C_L are written as follows (Klein et al. 2006):

$$C_L = C_{L0} + C_{L\alpha}\alpha + C_{Lq}\frac{q\bar{C}}{2U_0} + C_{L\delta e}\delta_e$$

$$C_m = C_{m0} + C_{m\alpha}\alpha + C_{mq}\frac{q\bar{C}}{2U_0} + C_{m\delta e}\delta_e$$

(3.1)

$C_{L(.)}$ and $C_{m(.)}$ are non-dimensional parameters which need to be extracted, and U_0 is the velocity at which the aircraft is trimming. The derivative of $C_{L\alpha}$ represents the lift curve slope and $C_{m\alpha}$ shows the static stability of an aircraft. C_{Lq} and C_{mq} are damping derivatives. $C_{L\delta_e}$ and $C_{m\delta_e}$ are representing the control effectiveness of an elevator deflection of the aircraft (Majeed 2014). The measurements of acceleration for the z-axis a_z and pitch $\dot{q}$ are given by Chowdhary and Jategaonkar (2010), Majeed and Kar (2013)

© The Author(s), under exclusive license to Springer Nature Singapore Pte Ltd. 2021   21
M. Mohamed and V. Dongare, *Aircraft Aerodynamic Parameter Estimation from Flight Data Using Neural Partial Differentiation*,
SpringerBriefs in Applied Sciences and Technology,
https://doi.org/10.1007/978-981-16-0104-0_3

Table 3.1 Mass, geometry, and inertia of aircraft (longitudinal dynamics)

Geometry	$\bar{c} = 1.904$ m (mean chord)
	$b = 14.7$ m (wing span)
	$S = 25.7$ m^2 (planform area)
Engines Inclination angle	$\sigma_t = 2$ deg
location of engine from CG	$l_{tx} = -4.3616$ m
total thrust	$l_{ty} = 0$ m
	$l_{tz} = -0.58985$ m
	$F_e = 7546$ N
Inertia mass	$I_x = 21562$ Kg.m^2
	$I_y = 79712$ Kg.m^2
	$I_z = 90861$ Kg.m^2
	$I_{xz} = 8056$ Kg.m^2
	$I_{xy} = I_{yz} = 0$ Kg.m^2
	$m = 5866.0$ Kg.m^2

$$a_z = -\frac{\bar{q}s}{m} C_L \cos\alpha - \frac{F_e}{m} \sin\sigma_t$$
$$\dot{q} = \frac{\bar{q}s\bar{c}}{I_y} C_m + \frac{pr}{I_y}(I_z - I_x) + (r^2 - p^2)\frac{I_{xy}}{I_y} + \frac{F_e}{I_y}(l_{tx}\sin\sigma_t + l_{tz}\cos\sigma_t) \quad (3.2)$$

where m is mass, $\bar{q}$ is dynamic pressure, s is planform area of wing, $\bar{c}$ is mean chord length, α is angle of attack (AoA), σ_t is engine inclination angle, F_e is total thrust force, p roll rate, r is yaw rate, $I_{(.)}$ is moment of inertia about an axis, and $l_{(.)}$ is the location of engine from the C.G. Their numerical value are tabulated in Table 3.1 (Majeed and Dongare 2015). The primary investigation of parameter estimation was carried out with simulated data of small transport aircraft.

3.2 Simulated Data Results for Longitudinal Dynamics

This section describes aircraft neural modeling and parameter estimation from simulated data. The flight simulation was carried out by the two elevator input δ_e, sequences of 3-2-1-1 for the time duration of 50 s. Moreover, the aircraft was trimmed at flight condition of the angle of attack 3.588 deg and Mach 0.25 at an altitude of 2800 m. These simulated data are at a frequency 40 Hz. Considering the input vector of α, q, δ_e and output vector of C_m and C_L to the neural network, it is able to establish the longitudinal dynamics of an aircraft neural model. The time histories of these signals are given in Fig. 3.1. Neural partial differentiation (NPD) method discussed in Chap. 2 is applied to the established neural model for the online estimation of aircraft aerodynamic derivatives. The estimation results using NPD are procured at the end of training of the neural network. Noise statistics of estimates, say standard deviation (SD) and relative standard deviation (RSD) are calculated independently after the training process of network (Sinha et al. 2013). Unlike delta and zero methods (Ghosh 2007), each of the aerodynamic derivatives of an aircraft are estimated

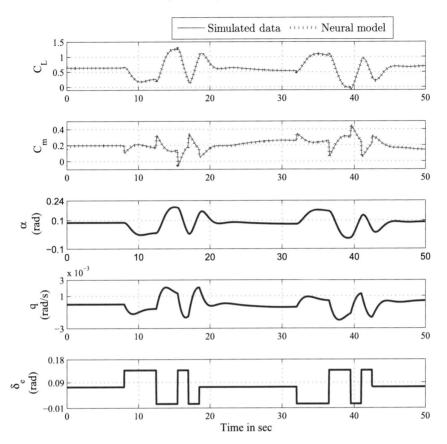

Fig. 3.1   Time history response of simulated data and neural model (longitudinal dynamics)

in real time by using NPD. Figure 3.2 shows the estimated values of the aerodynamic derivatives using the NPD method of simulated data with respect to the data points. It can be inferred from the plots that there is marginal variation in the aerodynamic derivatives with respect to the different data points. The variation of parameters with the change in the number of iterations is plotted in Fig. 3.3. The pitch acceleration $\dot{q}$ and normal acceleration (a_z) are reconstructed from a neural model, and they are matched well with those simulated data as shown in Fig. 3.4. This ensures that the estimation results for the stability and control derivatives are accurate. These derivatives are compared with the values obtained from OEM, and they are tabulated in Table 3.2. Since these estimated parameters from simulated data are very close to wind tunnel values, this motivates the application of NPD to flight data.

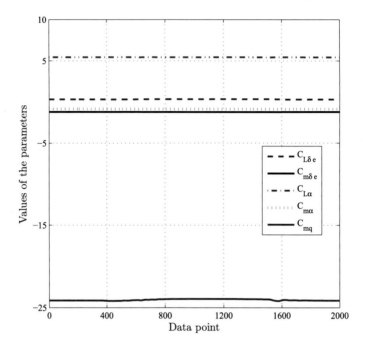

Fig. 3.2 Variation in parameters w.r.t. simulated data points (longitudinal dynamics)

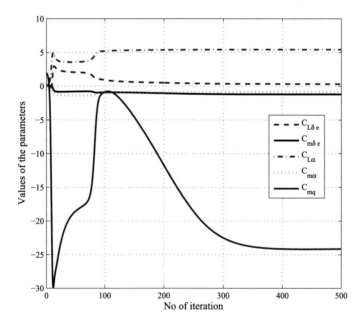

Fig. 3.3 Variation in parameters w.r.t. number of iterations during the training of simulated data (longitudinal dynamics)

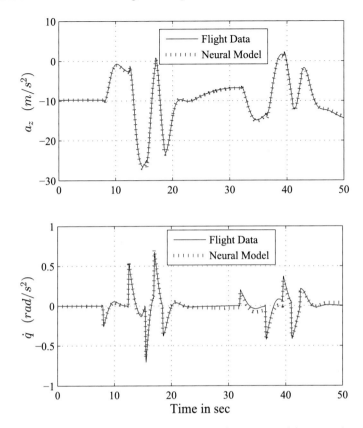

Fig. 3.4 Comparison of a_z and $\dot{q}$ from simulated data and reconstructed from neural model (longitudinal dynamics)

Table 3.2 Estimated aerodynamic derivatives from simulated data (longitudinal dynamics)

Parameter	Wind Tunnel	NPD	OEM
C_{L_α}	5.5338	5.438	4.5346
		(0.09)	(0.13)
$C_{L_{\delta e}}$	0.4318	0.3272	0.5923
		(5.0)	(0.12)
C_{m_α}	−1.1768	−0.853	−1.2586
		(0.27)	(0.02)
C_{m_q}	−23.0223	−24.13	−22.7156
		(0.025)	(1.84)
$C_{m_{\delta e}}$	−1.4168	−1.20	−1.3553
		(0.32)	(0.03)

*The values in parenthesis denote relative standard deviation values in percentage

3.3   Flight Data Results for Longitudinal Dynamics

This section describes the estimation of aerodynamic derivatives from flight data. For this, NPD method is applied to flight data of small transport aircraft. The flight testing was carried out at flight condition of the angle of attack 6.28 deg and Mach 0.22 at an altitude of 2729 m. The neural model of an aircraft system has been established by the training of input vector $[\alpha, q, \delta_e]$ and output vector $[C_L, C_m]$ to neural network (Majeed and Dongare 2015). Figure 3.5 shows that the flight data and the estimated responses for the longitudinal motion variables, and they are in close agreement. Figure 3.6 shows the estimated values of the aerodynamic derivatives using the neural partial differentiation method of flight data with respect to the data points. It can

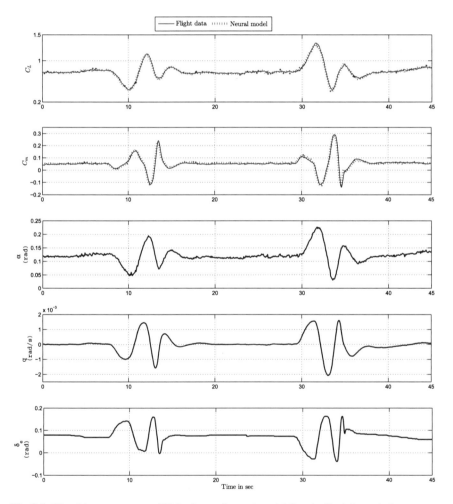

Fig. 3.5   Time history response of flight data and neural model (longitudinal dynamics)

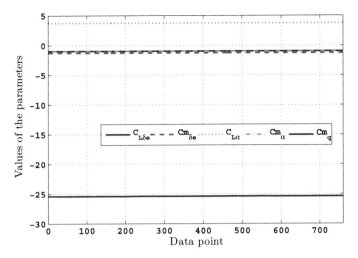

Fig. 3.6 Variation in parameters with respect to flight data points (longitudinal dynamics)

be inferred from the plots that there is a variation in some of the aerodynamic derivatives with respect to the different data points due to noise, otherwise the estimated parameters are closer to the wind tunnel values. The variation of parameters with the change in the number of iterations is plotted in Fig. 3.7. As the number of iterations increases, the parameters attain a stable value of their estimates. Figure 3.8 compares between $\dot{q}$ derived from flight data and from neural model. Normal acceleration (a_z) of flight data is compared with derived a_z from neural model, and shows a satisfactory match between them. Aircraft stability and control parameters are estimated by using NPD and OEM method from the flight data, and they are tabulated in Table 3.3.

3.4 Validation of Estimated Neural Model

The estimated aircraft neural model is needed to be verified with complementary flight data. For this, the neural network is trained with certain data set and new data set passed through the trained network. The output of networks is used to compute a_z for a given input of a complementary flight data, and compared with a_z derived from the lift force coefficient obtained from the wind tunnel value. The comparison plot of a_z signals is given in Fig. 3.9. The time history of estimated response a_z shows a mismatch with a_z derived from the wind tunnel while matches well with flight data. This reconfirms that estimates of aerodynamic derivatives are not closer to wind tunnel values as shown in Table 3.3 and the estimated neural model is valid.

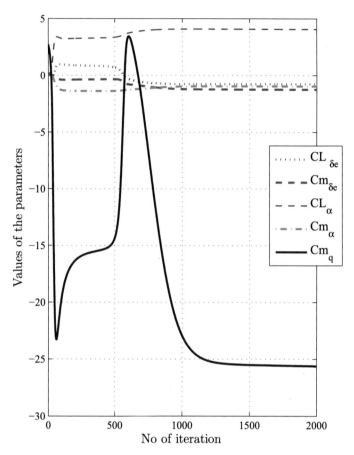

Fig. 3.7 Variation in parameters w.r.t. number of iterations during the training (longitudinal dynamics)

3.5 Summary

Neural partial differentiation method (NPD) is applied to simulated and flight data of small transport aircraft to estimate flight longitudinal stability and control derivatives. For this purpose, initially, a neural model of multi-input and multi-output (MIMO) aircraft system is established. The primary investigation of longitudinal parameter estimation is carried out from simulated data, and they found that estimates are very close to their wind tunnel values. NPD method is employed to extract the longitudinal aerodynamics parameter from flight data, and estimated parameters are comparable with estimates obtained from the output error method. Since the initial values of parameters are not available in a practical situation, the different neural network approach works well with flight data. Finally, the complementary flight data has been used to validate the identified neural model of aircraft.

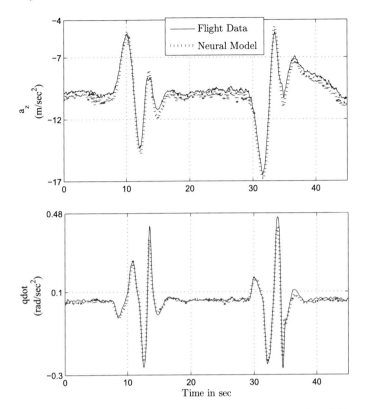

Fig. 3.8 Comparison of a_z and $\dot{q}$ from flight data and reconstructed from neural model (longitudinal dynamics)

Table 3.3 Estimated aerodynamic derivatives from flight data (longitudinal dynamics)

Parameter	Wind Tunnel	NPD	OEM
C_{L_α}	5.5338	4.783	4.5346
		(0.05)	(0.13)
$C_{L_{\delta_e}}$	0.4318	0.6272	0.5923
		(0.16)	(0.12)
C_{m_α}	−1.1768	−0.9455	−1.2586
		(0.09)	(0.02)
C_{m_q}	−23.0223	−25.34	−22.7156
		(0.025)	(1.84)
$C_{m_{\delta_e}}$	−1.4168	−1.2418	−1.3553
		(0.04)	(0.03)

*The values in parenthesis denote relative standard deviation values in percentage

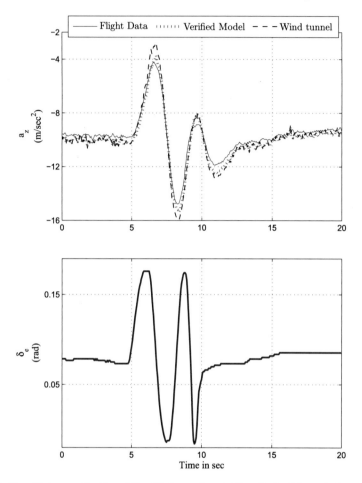

Fig. 3.9 Time history verification of identified neural model (longitudinal dynamics)

References

G. Chowdhary, R. Jategaonkar, Aerodynamic parameter estimation from flight data applying extended and unscented kalman filter. Aerosp. Sci. Technol. **14**(2), 106–117 (2010)

V. Klein, E.A. Moreli, *Aircraft System Identification Theory and Practice*, AIAA, Education Series (American Institute of Aeronautics and Astronautics, Reston, VA, 2006)

M. Majeed, *Aircraft System Identification and Control* (LAP LAMBERT Academic Publishing, Saarbrücken, 2014)

M. Majeed, V. Dongare, Neural partial differentiation based aircraft parameter estimation from flight data, in *Indian Control Conference* (2015), pp. 397–402

M. Majeed, I.N. Kar, Aerodynamic parameter estimation using adaptive unscented kalman filter. Int. J. Aircr. Eng. Aerosp. Technol. **85**(4), 267–279 (2013)

M. Sinha, R.A. Kuttieri, S. Chatterjee, Nonlinear and linear unstable aircraft parameter estimations using neural partial differentiation. J. Guid. Control Dynam. **36**(4), 1162–1176 (2013). https://doi.org/10.2514/1.57029

S. Singh, A.K. Ghosh, Estimation of lateral-directional parameters using neyral networks based modified delta method. Aeronaut. J. **111**(3150), 659–667 (2007)

Chapter 4
Identification of Aircraft Lateral-Directional Derivatives

The application of neural networks combined with partial differentiation of the neural outputs is discussed in this chapter to estimate lateral-directional flight stability and control derivatives from flight data. Primary investigation is carried out with simulated data and results are found to be encouraging to apply with flight data. Unlike longitudinal dynamics of aircraft, cross coupling parameters representing the stability and control of aircraft are presented in the lateral-directional dynamics of aircraft. As result, accurate estimation of such parameters is a challenging task referred to the lateral-directional dynamics of the complex aircraft system. Therefore, black box modeling technique of neural network is employed to model this complex system (Ghosh et al. 1998). The introduced method of NPD is used to extract cross coupling and other aerodynamic parameters describing the aircraft lateral-directional system dynamics (Singh and Ghosh 2007).

4.1 Lateral-Directional Aircraft Model

The differential equations in terms of motion-related variables say β angle of side slip, p roll rate, and r yaw rate are referred to the lateral-directional dynamics of aircraft. The time history of these variables is generated by the application of surface deflections δ_a and δ_r as aileron and rudder inputs, respectively. This aircraft dynamics are referred in terms of side force coefficient C_y, rolling moment coefficient C_l, and yawing moment coefficient C_n. These coefficients are represented by the Taylor series expansion as

$$
\begin{aligned}
C_y &= C_{y_0} + C_{y_\beta}\beta + C_{y_p}\frac{pb}{2U_0} + C_{y_r}\frac{rb}{2U_0} + C_{y_{\delta a}}\delta a + C_{y_{\delta r}}\delta r \\
C_l &= C_{l_0} + C_{l_\beta}\beta + C_{l_p}\frac{pb}{2U_0} + C_{l_r}\frac{rb}{2U_0} + C_{l_{\delta a}}\delta a + C_{l_{\delta r}}\delta r \\
C_n &= C_{n_0} + C_{n_\beta}\beta + C_{n_p}\frac{pb}{2U_0} + C_{n_r}\frac{rb}{2U_0} + C_{n_{\delta a}}\delta a + C_{n_{\delta r}}\delta r
\end{aligned}
\tag{4.1}
$$

© The Author(s), under exclusive license to Springer Nature Singapore Pte Ltd. 2021
M. Mohamed and V. Dongare, *Aircraft Aerodynamic Parameter Estimation from Flight Data Using Neural Partial Differentiation*,
SpringerBriefs in Applied Sciences and Technology,
https://doi.org/10.1007/978-981-16-0104-0_4

where $C_{y_{(.)}}$, $C_{l_{(.)}}$, and $C_{n_{(.)}}$ are non-dimensional parameter which need to be extracted, and U_0 velocity at which aircraft is trimming (Klein et al. 2006).

The static directional (Weathercock) stability of aircraft can be ensured by the positive value of C_{n_β}. The wings and the horizontal tailplane show a slightly positive effect on C_{n_β} while the fuselage causes C_{n_β} to decrease to a negative value. In order to compensate this effect, a vertical tailplane is normally used.

C_{l_β} is one of the most important parameters for lateral-directional stability and handling qualities. A negative value of C_{l_β} ensures the roll stability of aircraft.

During the rolling motion of aircraft, one wing of the aircraft goes up, while the other one goes down. This motion changes the effective angle of attack and thus also the lift of the wings. The upward going wing will get a lower lift, while the downward moving wing will experience a bigger amount of lift. The wing forces thus cause a moment opposite to the rolling motion. As a result, the rolling moment coefficient due to roll rate C_{l_p} is highly negative and shows a strong damping effect to rolling motion.

C_{l_p} is the damping in roll derivative and determines the roll substance. The change in rolling velocity causes change in rolling movement. The positive roll rate makes restoring moment and hence derivative negative value.

The side force coefficient C_{y_p} is contributed by the vertical tailplane. This vertical tailplane experiences an effective angle of attack due to its horizontal motion while rolling. This causes a horizontal force. The coefficient C_{y_p} is considered to be negative for a negative horizontal force toward a positive roll rate.

The yawing moment induced by roll rate has contributions from both the vertical tail and the wing, and it is represented by C_{n_p} derivative. This is a cross derivative and hence influences the Dutch-roll mode frequency.

Yaw rate stability: The derivatives are referred to be in yawing motion C_{y_r}, C_{l_r} and C_{n_r}. C_{n_r} is the most important derivative among them, and its contribution comes from the vertical tailplane. Yawing motion causes a horizontal force on the vertical tailplane. This force damps the yawing motion. Therefore, C_{n_r} has a negative value.

Since a positive deflation δa causes a negative rolling $C_{l_{\delta a}}$ is negative and represents the aileron deflection.

The derivatives with respect to rudder deflection are $C_{Y_{\delta r}}$, $C_{l_{\delta r}}$, $C_{n_{\delta r}}$. $C_{l_{\delta r}}$ the rudder surface deflection in rolling is positive. $C_{n_{\delta r}}$ is the desirable derivative that represent the rudder effectiveness.

Accelerometer measurements a_y, $\dot{p}$, and $\dot{r}$ are used in lateral-directional derivative estimation given as (Jategaonkar 2006)

$$
\begin{aligned}
a_y &= \tfrac{1}{g}\left[\tfrac{\bar{q}s}{m}C_y + (pq+\dot{r})X_{ay} - (r^2-p^2)Y_{ay} + (rq-\dot{p})Z_{ay}\right] \\
\dot{p} &= \tfrac{qsb}{I_xI_z-I_{xz}^2}\left[I_zC_l + I_{xz}C_n\right] - qr\tfrac{I_{xz}^2-I_yI_z+I_z^2}{I_xI_z-I_{xz}^2} + pq\tfrac{I_{xz}(I_x-I_y+I_z)}{I_xI_z-I_{xz}^2} \\
\dot{r} &= \tfrac{qsb}{I_xI_z-I_{xz}^2}\left[I_xC_n + I_{xz}C_l\right] + pq\tfrac{I_x^2-I_yI_z+I_z^2}{I_xI_z-I_{xz}^2} - qr\tfrac{I_{xz}(I_x-I_y+I_z)}{I_xI_z-I_{xz}^2}
\end{aligned}
\tag{4.2}
$$

Table 4.1 Mass, geometry, and inertia of aircraft (lateral-directional dynamics)

	$\bar{c} = 1.904\,\text{m}$ (mean chord)
	$b = 14.7\,\text{m}$ (wing span)
Geometry	$S = 25.7\,\text{m}^2$ (platform area)
Engines Inclination angle location of engine from CG Total Thrust	$\sigma_t = 2$ deg
	$l_{tx} = -4.3616\,\text{m}$
	$l_{ty} = 0$ m
	$l_{tz} = -0.58985\,\text{m}$
	$F_e = 7546$ N
	$I_x = 21562\ \text{Kg.m}^2$
	$I_y = 79712\ \text{Kg.m}^2$
Inertia	$I_z = 90861\ \text{Kg.m}^2$
	$I_{xz} = 8056\ \text{Kg.m}^2$
	$I_{xy} = I_{yz} = 0\ \text{Kg.m}^2$
Mass	$m = 5866.0\ \text{Kg.m}^2$
Accelerometer location from C.G. x, y and z direction	$X_{ay} = 0.98$
	$Y_{ay} = 0$
	$Z_{ay} = -0.31$

where m is mass, $\bar{q}$ is dynamic pressure, s is planform area of wing, b is the wing span, p roll rate, q is pitch rate, r is yaw rate, $I_{(.)}$ is moment of inertia about an axis, and $l_{(.)}$ is the location of engine from the C.G. Their numerical are tabulated in Table 4.1 (Dongare and Mohamed 2015). The primary investigation was carried out with simulated data of small transport aircraft.

4.2 Simulated Data Results for Lateral-Directional Dynamics

To estimate lateral-directional aerodynamic derivatives, aircraft responses pertaining to lateral-directional motion are generated at flight conditions of 4 deg angle of attack (AoA), Mach 0.24 at an altitude of 2600 m. For this, the doublet signals of aileron and rudder deflections are given as input to the aircraft system for the time duration of 50 s.

Figure 4.1 shows time histories of the input signals $\beta, p, r, \delta a$, and δa to the neural network and the output signals C_y, C_l, and C_n. The neural model of an aircraft system has been established by the training of input-output data. NPD can be applied to extract aerodynamic parameters from the neural model of aircraft, and their corresponding standard and relative standard deviations can be computed.

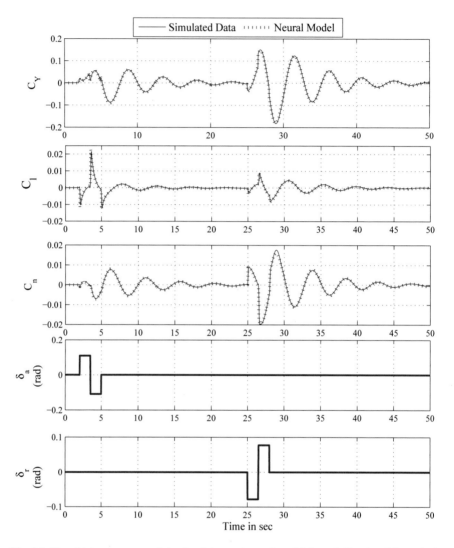

Fig. 4.1 Time history response of simulated data and neural model (lateral-directional dynamics)

Figure 4.2 shows the estimates of side force aerodynamic derivatives (parameter) using the NPD method for simulated data with respect to data points. It can be observed that there is a marginal variation in the aerodynamic derivatives with respect to the different data points. The variation of side force derivative with respect to the number of iteration is shown in Fig. 4.3. As the number of iterations increases, the parameters attain a stable value of its estimates. The close agreement of flight data with reconstructed a_y, $\dot{p}$, and $\dot{r}$ from the neural model in Fig. 4.4 indicates the accuracy of the estimated neural model of the aircraft system.

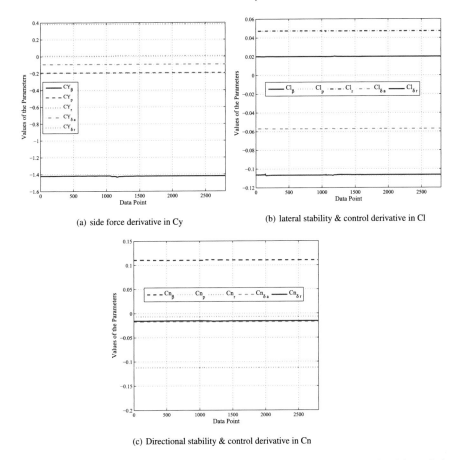

(a) side force derivative in Cy

(b) lateral stability & control derivative in Cl

(c) Directional stability & control derivative in Cn

Fig. 4.2 Variation in parameters with respect to simulated data points (lateral-directional dynamics)

The online estimation of lateral-directional aircraft stability and control parameter using OEM and NPD methods was achieved. Estimated aerodynamic derivatives from simulated data are tabulated in Table 4.2.

The estimated parameters from simulated data are very close to wind tunnel values results. This motivates the use of the proposed approach to flight data.

4.3 Flight Data Results for Lateral-Directional Dynamics

The flight test data of small transport aircraft were used to estimate side force, yawing, and rolling moment coefficients and demonstrated the efficacy of neural partial differentiation (NPD) method for online estimation of aircraft parameters

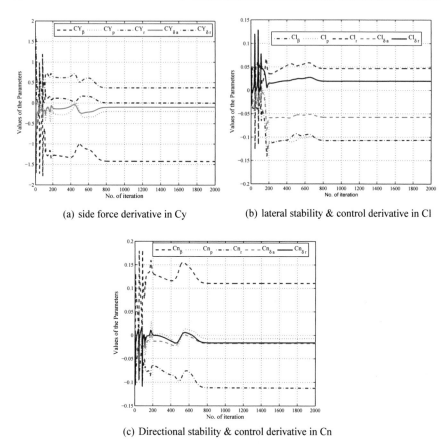

(a) side force derivative in Cy

(b) lateral stability & control derivative in Cl

(c) Directional stability & control derivative in Cn

Fig. 4.3 Variation in parameters w.r.t. number of iterations during training for simulate data (lateral-directional dynamics)

(Mohamed and Dongare 2018). During the flight test, the aircraft has trimmed at flight condition of 6.28 deg angle of attack, Mach 0.22 at an altitude of 2652 m.

The time history response of flight data and estimated responses are given in Fig. 4.5, and found that they are in close agreement with each other. This ensures that the dynamics of the aircraft model have been accurately identified.

The variation of parameters associated with side force, lateral stability, directional stability, and control with respect to data points are shown in Fig. 4.6. The variation of these parameters with the number of iterations is given in Fig. 4.7. This can be observed from Fig. 4.7 that certain parameter shows their variation due to the influence of noise. The rest of the parameters are closer to the wind tunnel values. The close agreement of flight data with reconstructed a_y, $\dot{p}$, and $\dot{r}$ from the neural model in Fig. 4.8 indicates the accuracy of the estimated neural model of the aircraft system. Estimated aerodynamic derivatives from flight data are tabulated in Table 4.3.

Fig. 4.4 Comparison of a_y, $\dot{p}$, and $\dot{r}$ from simulated data and reconstructed from Neural Networks (lateral-directional dynamics)

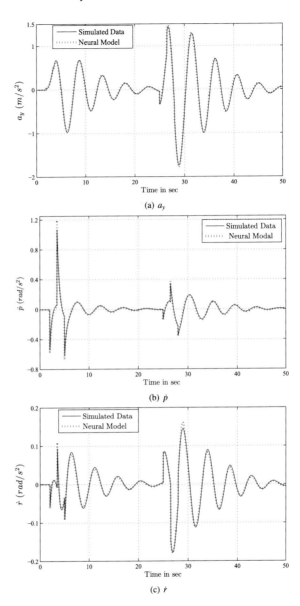

(a) a_y

(b) $\dot{p}$

(c) $\dot{r}$

Table 4.2 Estimated aerodynamic derivatives from simulated data (lateral-directional dynamics)

Parameters	True value	NPD	OEM
Cy_b	−1.43	−1.40 (0.1756)	−1.42 (0.49)
Cl_b	−0.109	−0.102 (0.1093)	−0.114 (1.02)
Cn_b	0.104	0.100 (0.31)	−0.114 (0.38)
Cy_p	−0.102	−0.125 (0.18)	−0.191 (19.56)
Cl_p	−0.599	−0.593 (0.07)	−0.597 (1.12)
Cn_p	−0.175	−0.189 (0.76)	−0.124 (2.04)
Cy_r	0.454	0.0454 (0.23)	1.97 (2.36)
Cl_r	0.220	0.297 (0.14)	0.365 (1.57)
Cn_r	−0.141	−0.388 (0.13)	−0.236 (1.13)
$Cy_{\delta a}$	−0.002	−0.20 (0.92)	−0.076 (8.76)
$Cl_{\delta a}$	−0.119	−0.107 (0.21)	−0.124 (0.85)
$Cn_{\delta a}$	−0.011	−0.007 (0.74)	−0.005 (8.95)
$Cy_{\delta r}$	0.328	0.389 (0.12)	0.248 (2.53)
$Cl_{\delta r}$	0.051	0.047 (0.42)	0.033 (2.02)
$Cn_{\delta r}$	−0.0875	−0.111 (0.38)	−0.078 (0.45)

*The values in parenthesis denote relative standard deviation values in percentage

4.4 Validation of Estimated Neural Model

The estimated neural model of lateral-directional aircraft dynamics is required to be validated with complementary flight data of the same aircraft at the same flight condition. This new set of complementary flight data is applied to the trained network of the estimated model as input signals. The output of the network gives the lateral acceleration a_y for the given input of flight data. The lateral acceleration a_y can also be computed by using the side force coefficient obtained either from the wind tunnel or the estimates of OEM. These computed lateral accelerations are compared with measured a_y in complementary flight data for the same input. Figure 4.9 illustrates the comparison of these accelerations, and a good match between these flight measured

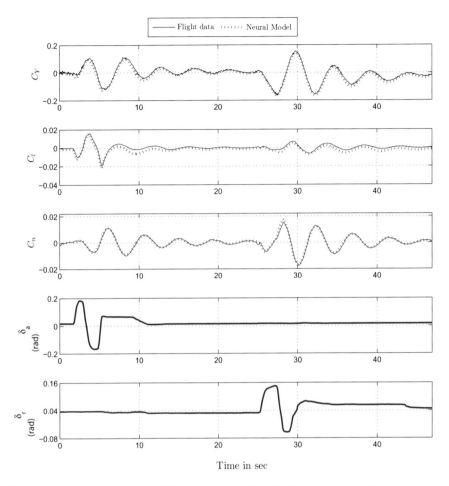

Fig. 4.5 Time history response of flight data and neural model (lateral-directional dynamics)

and predicted response is witnessed. Hence the estimated neural model is valid, and estimated derivatives using NPD are accurate to represent the aircraft dynamics.

4.5   Summary

The neural partial differentiation (NPD) method is employed to simulated and flight data of rigid aircraft to estimate lateral-directional flight stability and control parameters. For this purpose, initially neural model of multi-input and multi-output (MIMO) aircraft system is established. The primary investigation of lateral-directional parameter estimation is carried out from simulated data, and found that the estimates are

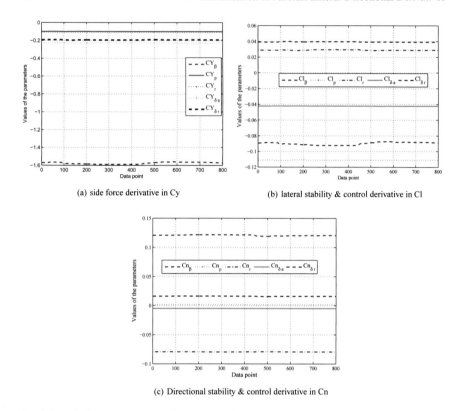

(a) side force derivative in Cy

(b) lateral stability & control derivative in Cl

(c) Directional stability & control derivative in Cn

Fig. 4.6 Variation in parameters with respect to flight data points (lateral-directional dynamics)

very close to wind tunnel values. The neural partial differentiation method is applied to extract the lateral-directional aerodynamic parameters from flight data, and the estimated parameter are comparable with estimates obtained from OEM. Since the initial values of parameters are not available in the practical situation as well as OEM requires these initial parameters, the neural network approach works well with the flight data. Finally, the identified neural model is validated by complementary flight data of small transport aircraft.

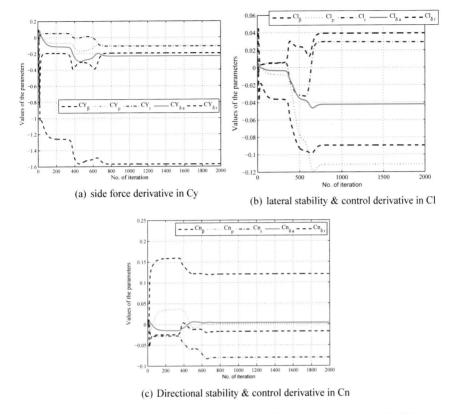

(a) side force derivative in Cy

(b) lateral stability & control derivative in Cl

(c) Directional stability & control derivative in Cn

Fig. 4.7 Variation in parameters with respect to number of iterations during training for flight data (lateral-directional dynamics)

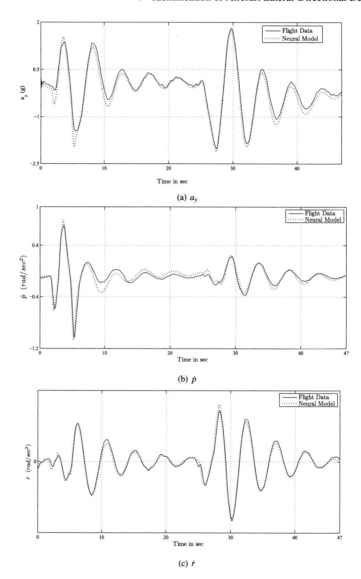

Fig. 4.8 Comparison of a_y, $\dot{p}$, and $\dot{r}$ from flight data and reconstructed from neural networks (lateral-directional dynamics)

Table 4.3 Estimated aerodynamic derivatives from flight data (lateral-directional dynamics)

Parameters	Wind tunnel value	NPD	OEM
Cy_b	−1.432	−1.648 (0.53)	−1.423 (0.49)
Cl_b	−0.109	−0.0895 (1.43)	−0.113 (1.02)
Cn_b	0.103	0.120 (0.56)	0.114 (0.38)
Cy_p	−0.102	−0.366 (0.28)	−0.190 (19.56)
Cl_p	−0.599	−0.462 (0.14)	−0.597 (1.12)
Cn_p	−0.175	−0.047 (3.25)	−0.124 (2.04)
Cy_r	0.454	0.405 (0.63)	1.975 (2.36)
Cl_r	0.221	0.02 (1.29)	0.365 (1.57)
Cn_r	−0.141	−0.371 (0.22)	−0.236 (1.13)
$Cy_{\delta a}$	−0.002	−0.037 (0.34)	−0.076 (8.76)
$Cl_{\delta a}$	−0.0119	−0.111 (0.19)	−0.124 (0.85)
$Cn_{\delta a}$	−0.012	−0.002 (0.83)	−0.005 (8.95)
$Cy_{\delta r}$	0.2918	0.365 (0.25)	0.248 (2.53)
$Cl_{\delta r}$	0.0385	0.029 (9.26)	0.033 (2.02)
$Cn_{\delta r}$	−0.0875	−0.079 (0.21)	−0.077 (0.45)

*The values in parenthesis denote relative standard deviation values in percentage

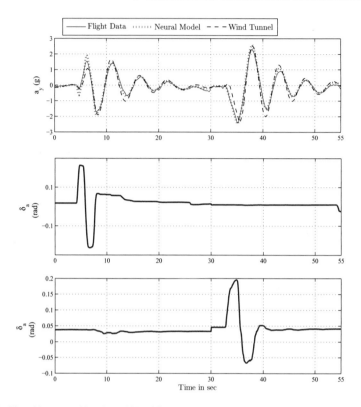

Fig. 4.9 Time history verification of identified neural model (lateral-directional dynamics)

References

V. Dongare, M. Mohamed, Lateral-directional aerodynamics parameter estimation using neural partial differentiation, in *2015 International Conference on Cognitive Computing and Information Processing (ccip)* (2015), pp. 1–6. https://doi.org/10.1109/CCIP.2015.7100730

A.K. Ghosh, S.C. Raisinghani, S. Khubchandani, Estimation of aircraft lateral-directional parameters using neural networks. J. Aircr. 35876–881 (1998)

R. Jategaonkar, Flight vehicle system identification: a time domain methodology. American Institute of Aeronautics and (2006). Astronautics

V. Klein, E.A. Moreli, *Aircraft System Identification Theory and Practice*, Education series (Reston, VAAIAA, 2006)

M. Mohamed, V. Dongare, Aircraft neural modeling and parameter estimation using neural partial differentiation. Aircr. Eng. Aerosp. Technol. **90**(5), 764–778 (2018). https://doi.org/10.1108/AEAT-02-2016-0021

S. Singh, A. Ghosh, Estimation of lateral-directional parameters using neural networks based modified delta method. Aeronat. J. **111**(3150), 659–667 (2007)

Chapter 5
Identification of a Flexible Aircraft Derivatives

This chapter discusses the neural modeling of the flexible aircraft and how to extract the aerodynamic derivatives and structural mode shape parameters of flexible modes using neural partial differentiation (NPD) method. The effects of flexibility on the flight dynamics of an aircraft have been shown to be quite significant, especially as the frequencies of its elastic modes become lower and approach those of the rigid body modes (Zerweckh et al. 1990; Meirovitch and Tuzcu 2001). The chapter discusses the neural modeling of the flexible aircraft and how to extract the aerodynamic derivatives and structural mode shape parameters of flexible modes using the NPD method. The effects of flexibility on the flight dynamics of an aircraft have been shown to be quite significant, especially as the frequencies of its elastic modes become lower and approach those of the rigid body modes (Colin et al. 2008; Majeed 2014). The characteristics of such flexible aircraft are altered significantly from those of a rigid aircraft, and the design of the flight control system may become drastically more complex (Bucharles and Vacher 2002). Therefore, mathematical modeling of a flexible aircraft for dynamic analysis and control system design is a major issue in flexible aircraft dynamics. The characteristics of such flexible aircraft are altered significantly from those of a rigid aircraft, and the design of the flight control system may become drastically more complex. Therefore, mathematical modeling of a flexible aircraft for dynamic analysis and control system design is a major issue in flexible aircraft dynamics.

5.1 Simulation of a Flexible Aircraft

In absence of flight data, simulated data of flexible aircraft is generated by using the postulated model. The simulated data contains the two elastic modes that get

M. Mohamed and V. Dongare, *Aircraft Aerodynamic Parameter Estimation from Flight Data Using Neural Partial Differentiation*,
SpringerBriefs in Applied Sciences and Technology,
https://doi.org/10.1007/978-981-16-0104-0_5

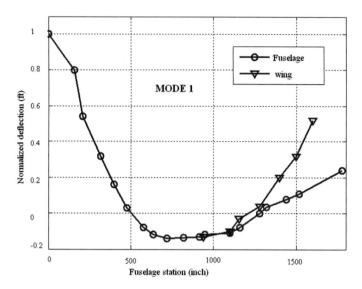

Fig. 5.1 Mode shape of symmetric modes indicating fuselage bending with wing participating in phase (flexible aircraft)

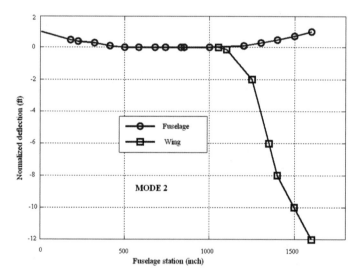

Fig. 5.2 Mode shape of antisymmetric modes indicating the wing bending and fuselage participating out of phase

excited in the longitudinal axis. The included two structural modes in flight simulation characterize (i) fuselage bending with wing participating in phase and (ii) wing bending and fuselage participating out of phase (Majeed et al. 2012). The mode shapes of these two modes of aircraft are given in Figs. 5.1 and 5.2.

Table 5.1 Mass, geometry, and inertia of flexible aircraft

Geometry	$\bar{c} = 4.664$ m (mean chord)
	$b = 21.336$ m (wing span)
	$S = 180.79$ m^2 (platform area)
	$\Lambda = 65$ deg (sweep angle)
Weight	$W = 130642.3$ Kg (net weight)
Inertia	$I_x = 1288066$ Kg.m^2
	$I_y = 8677503$ Kg.m^2
	$I_z = 9626605$ Kg.m^2
	$I_{xz} = -71453$ Kg.m^2
	$I_{xy} = I_{yz} = 0$ Kg.m^2
Modal damping	$\xi_1 = 0.02, \xi_2 = 0.02$
Modal generalizes	$M_1 = 248.94$ Kg.m^2
mass	$M_2 = 12998.0$ Kg.m^2

Mass, geometry, and inertia of flexible aircraft are shown in Table 5.1.

Postulated model of a flexible aircraft for pitching motion can be approximated by neglecting variations in velocity as given in Majeed (2014).

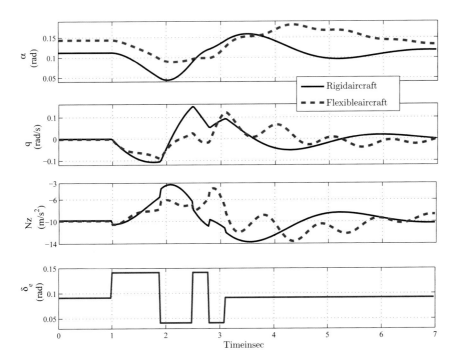

Fig. 5.3 Comparison of aircraft responses

$$\dot{\alpha} = q + \frac{\rho u S}{2M} C_Z$$
$$\dot{q} = \frac{\rho u^2 S \bar{c}}{2 I_y} C_m$$
(5.1)

Where C_Z and C_m represent the aerodynamic coefficients that consist of aerodynamic and structural derivatives. Air density ρ, total inertial velocity u, wing area S, wing chord $\bar{c}$, aircraft mass M, and the moment inertia I_y about y -axis are the other quantities used in the above equations. The aerodynamic coefficients C_Z and C_m used in aerodynamic models can be written as:

$$C_Z = C_{Z_0} + C_{Z_\alpha}\alpha + C_{Z_q}\frac{q\bar{c}}{2u} + C_{Z_{\delta e}}\delta e + C_{Z_{\eta 1}}\eta 1 + C_{Z_{\eta 2}}\eta 2 + \frac{\bar{c}}{2u}C_{Z_{\dot{\eta}1}}\dot{\eta}1 + \frac{\bar{c}}{2u}C_{Z_{\dot{\eta}2}}\dot{\eta}2$$
$$C_m = C_{m_0} + C_{m_\alpha}\alpha + C_{m_q}\frac{q\bar{c}}{2u} + C_{m_{\delta e}}\delta e + C_{m_{\eta 1}}\eta 1 + C_{m_{\eta 2}}\eta 2 + \frac{\bar{c}}{2u}C_{m_{\dot{\eta}1}}\dot{\eta}1 + \frac{\bar{c}}{2u}C_{m_{\dot{\eta}2}}\dot{\eta}2$$
(5.2)

We have to consider the elastic states of the flexible aircraft (η_1, η_2) and ($\dot{\eta}_1$, $\dot{\eta}_2$) to augment with rigid body dynamic model. For this, the generalized coordinates satisfying Eq. (5.2) is as follows:

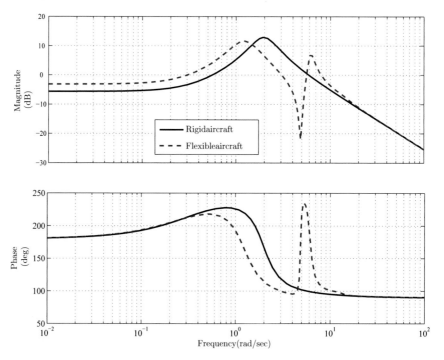

Fig. 5.4 Frequency response for pitch rate to elevator input ($\frac{q}{\delta e}$)

$$\ddot{\eta}_1 + 2\xi_1\omega_1\dot{\eta}_1 + \omega_1{}^2\eta_1 = \frac{\rho u^2 S \bar{c}}{2M_1} \left[C_\alpha^{\eta_1}\alpha + C_q^{\eta_1}\frac{q\bar{c}}{2u} + C_{\delta_e}^{\eta_1}\delta_e + \begin{pmatrix} C_{\eta 1}^{\eta_1}\eta_1 + C_{\eta 2}^{\eta_1}\eta_2 \\ +C_{\dot{\eta}1}^{\eta_1}\frac{\dot{\eta}_1\bar{c}}{2u} + C_{\dot{\eta}2}^{\eta_1}\frac{\dot{\eta}_2\bar{c}}{2u} \end{pmatrix} \right]$$

$$\ddot{\eta}_2 + 2\xi_2\omega_2\dot{\eta}_2 + \omega_2{}^2\eta_2 = \frac{\rho u^2 S \bar{c}}{2M_2} \left[C_\alpha^{\eta_2}\alpha + C_q^{\eta_2}\frac{q\bar{c}}{2u} + C_{\delta_e}^{\eta_2}\delta_e + \begin{pmatrix} C_{\eta 1}^{\eta_2}\eta_1 + C_{\eta 2}^{\eta_2}\eta_2 \\ +C_{\dot{\eta}1}^{\eta_2}\frac{\dot{\eta}_2\bar{c}}{2u} + C_{\dot{\eta}2}^{\eta_2}\frac{\dot{\eta}_2\bar{c}}{2u} \end{pmatrix} \right]$$

$$(5.3)$$

The simulated data for α, q, N_z (normal acceleration) and the elastic states η_1, η_2, $\dot{\eta}_1$, $\dot{\eta}_2$ are generated. The aeroelastic effects would affect the aircraft response for a given control input. Simulated data plotted for rigid body and flexible aircraft is given in Fig. 5.3. Figure 5.3 shows that the α *and* q responses for a flexible aircraft differ quantitatively but also qualitatively from the rigid body responses. The frequency responses of the rigid model and the aeroelastic models are given in Fig. 5.4. Review of this result reveals that the two elastic modes are excited at frequencies of 6.29 rad/sec and 7.21 rad/sec with damping factor $\xi_1 = \xi_1 = 0.02$ (Majeed and Jatinder 2013; Mohamed 2017). The rigid body model would lead to erroneous and unacceptable results for the flexible aircraft.

5.2 Parameter Estimation Results and Discussion for a Flexible Aircraft

This section explains the extracting of rigid body and elastic body derivatives from a neural model of flexible aircraft. For this, neural model of flexible aircraft system was established by training of network with input vector of $[Cz, Cm, \ddot{\eta}_1, \ddot{\eta}_2]$ and output vector of $[\delta e, \alpha, q, \dot{\eta}_1, \dot{\eta}_2, \eta_1, \eta_2]$. NPD method was used for the estimation of flexible aircraft parameters. The analysis was carried out with simulated flight data of flexible aircraft. NPD can be applied to extract flexible aircraft parameters from the neural model of aircraft, and their corresponding standard and relative standard deviations can be computed. Figure 5.5 shows time histories of the output signals ($Cz, Cm, \ddot{\eta}_1, \ddot{\eta}_2$) to the neural network and Fig. 5.6 shows input signals ($\delta e, \alpha, q, \dot{\eta}_1, \dot{\eta}_1, \eta_2, \eta_2$). The flight simulation has been carried out by input frequency sweep signals of elevator for time duration of 90 s.

Figure 5.7 shows the estimation of parameters Cm_α, $C_{\eta_1}^{\eta_1}$, $C_{\delta e}^{\eta_1}$, ω_1 using the NPD method for simulated data with respect to data points. It shows there is a marginal variation in $C_{\delta e}^{\eta_1}$ with respect to the different data points.

The variation of parameters Cm_α, $C_{\eta_1}^{\eta_1}$, $C_{\delta e}^{\eta_1}$, ω_1 with respect to the number of iteration is shown in Fig. 5.8. As the number of iteration increases, the parameters attain constant value of its estimates.

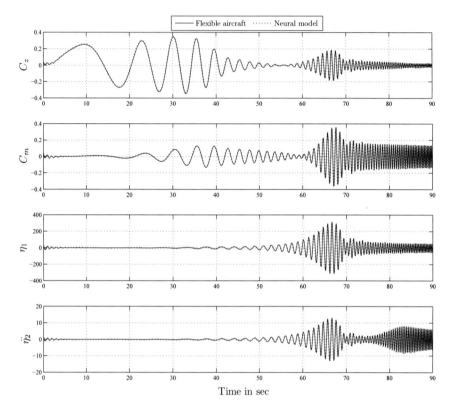

Fig. 5.5 Time history response of flexible aircraft and neural model (output signal)

The time history response of simulated data and estimated responses are given in Fig. 5.5, and found that they are in close agreement with each other. This ensures that the dynamics of the flexible aircraft model have been accurately identified.

The comparison plot between the time response of N_z from simulated data and reconstructed from aircraft neural model is shown in Fig. 5.9.

Besides the estimation of the stability and control derivative of flexible aircraft, the structural mode shape parameters are also estimated from the simulated data, and they are tabulated in Table 5.2.

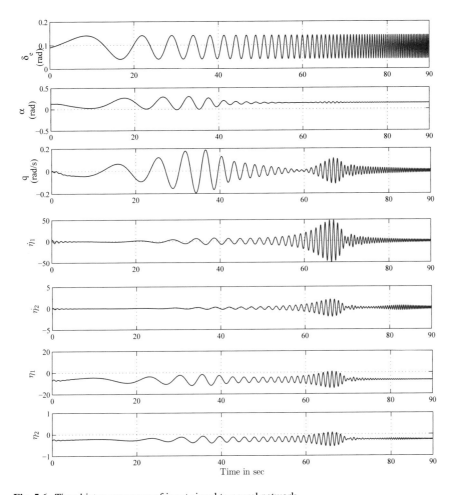

Fig. 5.6 Time history response of input signal to neural network

5.3 Validation of Estimated Neural Model

Estimated neural model of flexible aircraft is required to be verified with complementary data set simulated data. For this, the neural network is trained with elevator input signal of frequency sweep, and then a new data set of elevator input signal 3-2-1-1 is passed through the trained network. The output of the network is used to compute N_z for the given input data. The time responses of the normal accelerations N_z obtained from simulated data and neural model are given in Fig. 5.10. These responses of N_z are well matched with each other and thus estimated. The estimated model of flexible aircraft is valid.

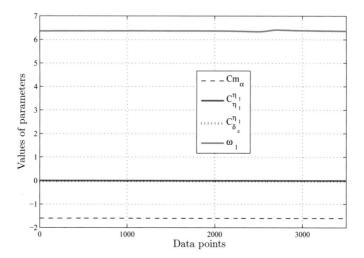

Fig. 5.7 Variation in parameters w.r.t. data points of flexible aircraft

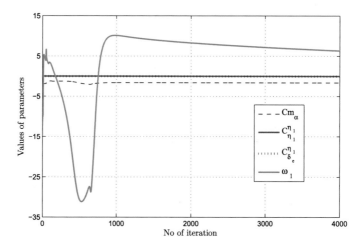

Fig. 5.8 Variation in parameters w.r.t. number of iterations during training of flexible aircraft

5.4   Summary

Neural partial differentiation (NPD) is applied to simulated data of flexible aircraft to extract aerodynamic derivatives and frequencies of the structural modes of the flexible aircraft. The identified neural model of aircraft is validated by complimentary simulated data set of flexible aircraft. For the parameter estimation purpose, primarily neural model of multi-input and multi-output (MIMO) flexible aircraft is established. Flight simulation of flexible aircraft is carried out for an elevator input of frequency sweep. Moreover, symmetric and antisymmetric structural modes are included in

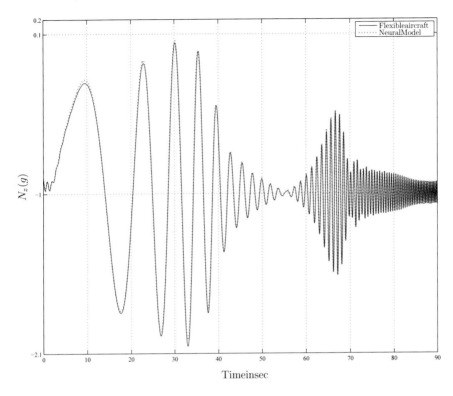

Fig. 5.9 Comparison of N_z of flexible aircraft

the flight simulation to represent the flexibility of aircraft. Unlike the delta or zero method of neural networks, the neural partial differentiation (NPD) is able to estimate aerodynamic parameters directly from the neural model with their relative standard deviations. The estimated parameters are comparable with those obtained from output error method, which is used for flight simulation. Since the initial values of certain aerodynamic parameters of flexible aircraft are not available in a practical situation, it is proposed that the approach of neural partial differentiation works well.

Table 5.2 Estimated aerodynamic derivatives of flexible aircraft

Parameter	True value	NPD	OEM
$Cz_{\delta e}$	−0.435	−0.429 (6.34)	−0439 (2.23)
Cz_α	−2.922	−2.744 (0.09)	−2.963 (0.29)
Cz_q	14.765	15.545 (1.69)	14.56 (4.19)
$Cz_{\dot\eta_1}$	−0.0848	−0.052 (1.67)	–
$Cz_{\dot\eta_2}$	1.03	1.087 (1.95)	–
Cz_{η_1}	−0.0288	−0.029 (0.17)	−0.0291 (0.54)
Cz_{η_2}	0.306	0.31 (0.51)	–
$Cm_{\delta e}$	−2.578	−2.293 (0.28)	−2.546 (0.31)
Cm_α	−1.66	−1.566 (0.17)	−1.639 (0.56)
Cm_q	−34.75	−32.476 (1.47)	−34.57 (0.30)
$Cm_{\dot\eta_1}$	−0.159	−0.112 (0.92)	0.157(1.37)
$Cm_{\dot\eta_2}$	1.23	0.628 (3.22)	–
Cm_{η_1}	−0.0321	−0.034 (0.13)	−0.031 (0.96)
Cm_{η_2}	0.025	0.027 (2.11)	–
$C_{\delta e}^{\eta_1}$	−0.0128	−0.013 (1.04)	−0.0127 (0.07)
$C_\alpha^{\eta_1}$	−0.0149	−0.016 (0.24)	−0.015 (0.12)
$C_q^{\eta_1}$	−0.095	−0.084 (5.78)	–
ω_1	6.29	6.254 (2.47)	–
ω_2	7.21	7.18 (3.47)	–
$C_{\dot\eta_2}^{\eta_1}$	−2e-04	2.2e-4 (6.14)	–
$C_{\eta_1}^{\eta_1}$	6e-05	6e-05 (0.13)	5.7e-05 (0.6)
$C_{\eta_2}^{\eta_1}$	9e-05	7e-05 (1.86)	–
$C_{\delta e}^{\eta_2}$	−0.064	−0.065 (1.66)	−0.0634 (0.19)
$C_\alpha^{\eta_2}$	0.026	0.026 (0.75)	–
$C_q^{\eta_2}$	0.012	−0.012 (4.57)	–
$C_{\dot\eta_1}^{\eta_2}$	0.009	0.009 (1.46)	–
$C_{\dot\eta_2}^{\eta_2}$	−0.298	−0.290 (0.54)	–
$C_{\eta_1}^{\eta_2}$	0.004	0.004 (0.09)	–

*The values in parenthesis denote relative standard deviation values in percentage

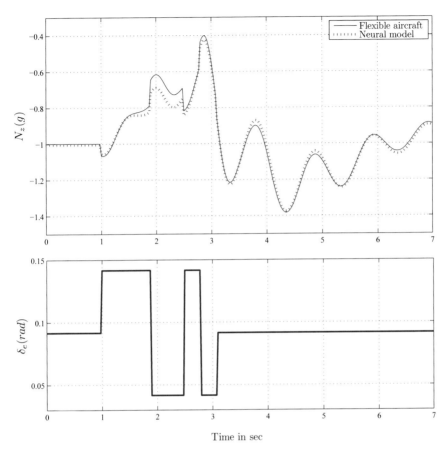

Fig. 5.10 Time history verification of identified neural model (flexible aircraft)

References

A. Bucharles, P. Vacher, Flexible aircraft model identification for control law design. Aerosp. Sci. Technol. 6591–598 (2002)

T. Colin, I. Christina, T. Mark, F. Edmund, N. Randall, R. Heather, *System Identification of Large Flexible Transport Aircraft* (American Institute of Aeronautics and Astronautics, AIAA Atmospheric Flight Mechanics Conference and Exhibit, 2008). https://doi.org/10.2514/6.2008-6894

M. Majeed, Parameter identification of flexible aircraft using frequency domain output error approach, in *International Conference on Advances in Control and Optimization of Dynamical Systems (ACODS)* (2014)

M. Majeed, S. Jatinder, Frequency and time domain recursive parameter estimation for a flexible aircraft, in *19th IFAC Symposium on Automatic Control in Aerospace* (2013), pp. 443–448

M. Majeed, I.N. Kar Singh, Identification of aerodynamic derivatives of a flexible aircraft. J. Aircraft **49**(2), 654–658 (2012). https://doi.org/10.2514/i.c031318

L. Meirovitch, I. Tuzcu, Multidisciplinary approach to the modeling of flexible aircraft, in *International Forum on Teroelasticity and Structural Dynamics* (2001), pp. 435–448

M. Mohamed, System identification of flexible aircraft in frequency domain. Aircraft Eng. Aerosp. Technol. **89**(6), 826–834 (2017). https://doi.org/10.1108/AEAT-09-2015-0214

S.H. Zerweckh, A.H.V. Flotow, J.E. Murray, Flight testing of highly flexible aircraft. J. Aircraft **27** (1990)

Chapter 6
Conclusions and Future Work

Neural partial differentiation (NPD) method is used for the parameter estimation. For this purpose initially, a neural model of multi-input and multi-output (MIMO) aircraft system is established. In addition to that, separate consideration of the longitudinal and lateral-directional dynamics simplifies the model considerably in terms of structure and the number of parameters. The primary investigation of aerodynamics parameter estimation is carried out from simulated data, and it is found that the estimates are very close to wind tunnel values. NPD is employed to extract the aerodynamic parameters from flight data. Further, the NPD method is used for flexible aircraft parameter estimation purpose, for that flight simulation of flexible aircraft is carried out for an elevator input of frequency sweep. Moreover, symmetric and antisymmetric structural modes are included in flight simulation to represent the flexibility of aircraft. The estimated parameters are comparable with estimates obtained from the output error method (OEM). Since the initial values of parameters are not available in practical situations as well as OEM requires these initial parameters, the neural network approach works well with the flight data. Unlike the delta or zero method of neural networks, the NPD is able to estimate aerodynamic parameters directly from the neural model with their relative standard deviation. The proposed neural network approach of neural partial differentiation works well for the parameter estimation. Finally, the identified models are validated by complementary flight data of aircraft.

© The Author(s), under exclusive license to Springer Nature Singapore Pte Ltd. 2021   59
M. Mohamed and V. Dongare, *Aircraft Aerodynamic Parameter Estimation from Flight Data Using Neural Partial Differentiation*,
SpringerBriefs in Applied Sciences and Technology,
https://doi.org/10.1007/978-981-16-0104-0_6

6.1   Summary

In the light of the above discussion, a summary of the monograph is as follows:

- Established a neural model of rigid aircraft from flight data and extracted its longitudinal and lateral-directional derivatives using neural partial differentiation method. The extracted derivatives are compared with the estimates obtained from OEM.
- The estimated neural model of rigid aircraft is validated by a complementary set of flight data. The results are encouraging and this method is applied to flexible aircraft that contain more number of unknown parameters.
- Besides the estimations of aerodynamic derivatives of flexible aircraft, unknown frequencies of structural modes of flexibility are also identified from its simulated data of frequency sweep input. The identified flexible aircraft model has been validated with a complimentary set of data that has been generated with the use of 3-2-1-1 input signal of an elevator.

6.2   Further Research

The use of neural partial differentiation is possible to extend for the following purposes:

- Performance analysis of Aircraft: This can be done by estimating the value of C_{D0} and C_L from the flight data obtained through the Roller Coaster maneuver of an aircraft. These terms are nonlinearly related as $C_D = C_{D0} + \frac{1}{\pi e AR} C_L^2$ for the drag polar characteristics of an aircraft.
- Extracting the parameters of dynamic systems such as underwater robotics which are appearing nonlinear to the states of system.

The solution of the parameter estimation can be derived from either EEM or NPD method. In the presence of noise in the signal, EEM-based estimated parameters are biased to those noises. Whereas the NPD method gives unbiased estimates with their noise statistics of std and relative std. This will introduce the novel estimation algorithm, which is the combination of NPD and EEM for the better prediction of system parameters.

Appendix A
Neural-Network-Based Solution of Ordinary Differential Equation

The function approximate of the physical system can be done by the neural network. The solution of ODE/PDE can be found out using the neural network Lagaris et al. (1998); Fojdl & Brause (2008); Hornik et al. (1989). First-order ordinary differential equation (ODE) can be written as $\frac{\partial y}{\partial x} = f(x, y)$ and the initial conditions (IC) $x = [0, 1]$. A trial solution is

$$y_T(x) = A + xN(x, \mathbf{o}) \tag{A.1}$$

where $N(x, \mathbf{o})$ is the output of a feed forward neural network with one input unit for x and weights $\mathbf{o}$. Note that satisfies the IC by construction. The error quantity to be minimized is given by

$$E[\mathbf{o}] = \sum_i \left\{ \frac{\partial y_T(x_i)}{\partial x} - f(x_i, y_T(x_i)) \right\}^2 \tag{A.2}$$

where the ODE's are points in $[0, 1]$. Since it is straightforward to compute the gradient of the error with respect to the parameters(coefficient). The same holds for all subsequent model problems.

$$\frac{\partial^2 y}{\partial x^2} = f(x, y, \frac{\partial y}{\partial x}) \tag{A.3}$$

For the initial value problem: $y(0) = A$ and $\frac{\partial y(0)}{\partial x} = A'$, the trial solution can be cast as

$$\phi_t(x) = A + A' + x^2 N(x, \mathbf{o}) \tag{A.4}$$

For the two point Dirichlet BC: $x(0) = A$ and $x(1) = B$, the trial solution is written as

$$y_T(x) = A(1 - x) + xB + x(1 - x)N(x, \mathbf{o}) \tag{A.5}$$

M. Mohamed and V. Dongare, *Aircraft Aerodynamic Parameter Estimation from Flight Data Using Neural Partial Differentiation*,
SpringerBriefs in Applied Sciences and Technology,
https://doi.org/10.1007/978-981-16-0104-0

In the above two cases of second-order ODE's, the error function to be minimized is given by the following equation:

$$E[\mathbf{o}] = \sum_i \left\{ \frac{\partial^2 y_T(x_i)}{\partial t^2} - f(x_i, y_T(x_i), \frac{\partial y_T(x_i)}{\partial x}) \right\}^2 \qquad (A.6)$$

References

J. Fojdl, R.W. Brause, The performance of approximating ordinary differential equations by neural nets, in *20th IEEE International Conference on Tools with Artificial Intelligence, 2008. ictai '08.*, vol. 2 (2008), pp. 457–464.-https://doi.org/10.1109/ictai.2008.44

K. Hornik, M. Stinchcombe, H. White, Multi layer feed forward neural networks are universal approximators. Neural Netw. **2**, 359–366 (1989).

I.E. Lagaris, A. Likas, D.I. Fotiadis, Artificial neural networks for solving ordinary and partial differential equations. IEEE Trans. Neural Netw. **9**(5), 987–100 (1998).

Appendix B
Output Error Method

In the output error method (OEM), the unknown parameters are obtained by minimizing the sum of weighted square differences between the measured outputs and model outputs. The estimation problem is nonlinear because of unknown parameter appears in the aircraft equations of motion and they are integrated to compute the states. Outputs are computed from states, control input, and parameters using the measurement equation. Iterative nonlinear optimization techniques are required to solve this nonlinear estimation problem Joao et al. (2005), Majeed and Kar (2011), Mehra (1970), Majeed (2014).

The mathematical model aircraft is assumed to describe the following general linear dynamics system representation.

$$
\begin{aligned}
&\dot{x}(t) = Ax(t) + Bu(t), x(t_0) = x_0 \\
&y(t) = Cx(t) + Du(t) \\
&z(t_k) = y(t_k) + v(t_k), k = 1, 2, 3, ...N
\end{aligned}
\tag{B.1}
$$

where x is the ($n_x + 1$) state variables, u the ($n_u + 1$) control input vector, y the ($n_z + 1$) system output vector, and measurement vector z is sampled at N discrete points. The Matrices A, B, C, and D contain the unknown system parameters and are given by

$$
\begin{aligned}
\Theta = [(A_{ij}, i = 1 \text{ to } n_x; j = 1 \text{ to } n_x)^T (B_{ij}, i = 1 \text{ to } n_x; j = 1 \text{ to } n_x)^T \\
(C_{ij}, i = 1 \text{ to } n_y; j = 1, n_x)^T (D_{ij}, i = 1 \text{ to } n_y; j = 1 \text{ to } n_u)]^T
\end{aligned}
\tag{B.2}
$$

In order to estimate the likelihood function to estimate the parameter of dynamic system represented in Eq. (B.1), the following assumptions are used:

- The exogenous input sequence $[u(t_k), k = 1, 2, 3...N]$ is independent of the system output.

© The Author(s), under exclusive license to Springer Nature Singapore Pte Ltd. 2021   63
M. Mohamed and V. Dongare, *Aircraft Aerodynamic Parameter Estimation from Flight Data Using Neural Partial Differentiation*,
SpringerBriefs in Applied Sciences and Technology,
https://doi.org/10.1007/978-981-16-0104-0

- The measurement errors $[v(t_k) = z(t_k) - y(t_k)]$ at different discrete points are statically independent, and are assumed to be distributed with zero means and covariance matrix R, that is, $E(v(t_k) = 0, E[v(t_k)v^T(t_l)] = R\delta_{kl}$.
- The system is corrupted by measurement noise only.
- Control inputs $u(t_k)$ are sufficiently and adequately (i.e., in magnitude and frequency) varied to excite directly or indirectly the various modes of the dynamics system being analyzed.

The maximum likelihood output error estimates of unknown parameters are obtained by minimizing the negative logarithm of the likelihood function. Figure B.1 shows a block schematic of the output error method (OEM). The cost function of this method is considered in (B.3).

$$J(\Theta) = \frac{1}{2} \sum_{k=1}^{N} [z(t_k) - y(t_k)]^T R^{-1} [z(t_k) - y(t_k)] + \frac{N}{2} \ln |R| \qquad (B.3)$$

where covariances matrix of the residuals and estimates can be obtained from the (B.4). When started from suitably specified initial valus, the estimates are iteratively updated using the Gauss-Newton method.

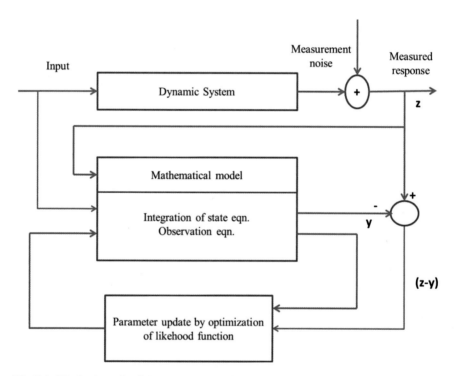

Fig. B.1 Block schematic of the output error method

$$R = \frac{1}{N} \sum_{k=1}^{N} [z(t_k) - y(t_k)]^T [z(t_k) - y(t_k)] \tag{B.4}$$

The algorithmic steps of OEM are given below

Give the initial value of the Θ, i.e., Θ_0. It may also consist of the initial value of states x_0 if not known and biases in measurements Δz if required.

Step 1: Set iteration $= 1$

Step 2: Compute the response and cost function J:

$$\dot{x}(t) = Ax(t) + Bu(t)$$
$$y(t) = Cx(t) + Du(t)$$
$$R = \frac{1}{N} \sum_{k=1}^{N} [z(t_k) - y(t_k)]^T [z(t_k) - y(t_k)] \tag{B.5}$$
$$J = \frac{1}{2} \sum_{k=1}^{N} [z(t_k) - y(t_k)]^T R^{-1} [z(t_k) - y(t_k)] + \frac{N}{2} \ln |R|$$

Step 3: Perturb the parameter j, i.e., Θ_j to $\Theta_i + \Delta\Theta_i$, so that system matrices becomes $A_p B_p C_p D_p$

Step 4: Compute the perturbation responses and update on Θ.

$$\begin{cases} \dot{x}_p(t) = A_p x_p(t) + B_p u(t) \\ y_p(t) = C_p x_p(t) + D_p u(t) \\ \frac{\partial y(t)}{\partial \Theta_j} = \frac{[y_p(t) - y(t)]}{\Delta \Theta_j} \\ \Delta_\Theta J(\Theta) = \sum_{k=1}^{N} \left[\frac{\partial y(t_k)}{\partial \Theta_j} \right]^T R^{-1} [z(t_k) - y(t_k)] \\ \Delta_\Theta^2 J(\Theta) = \sum_{k=1}^{N} \left[\frac{\partial y(t_k)}{\partial \Theta_j} \right]^T R^{-1} \left[\frac{\partial y(t_k)}{\partial \Theta_j} \right] \\ \Theta = \Theta + \left[\Delta_\Theta^2 J(\Theta) \right]^{-1} [\Delta_\Theta J(\Theta)] \end{cases} \tag{B.6}$$

Step 5: Increment the iteration count and jump back to step 2 to continue until the cost function reduces to zero approximately. Thus estimated parameter Θ is the updated at which cost function is minimized.

References

O. Joao, Q.P. Chu, J.A. Mulder, H.M. Balini, N. K., W. Vos, G. M. Output error method and two step method for aerodynamic model identification, in*AIAA Guidance, Navigation, and Control Conference and Exhibit*. (American Institute of Aeronautics and Astronautics, 2005). https://doi.org/10.2514/6.2005-6440

M. Majeed, Parameter identification of flexible aircraft using frequency domain output error approach, in *International Conference on Advances in Control and Optimization of Dynamical Systems (acods)* (2014)

M. Majeed, I.N. Kar, Identification of aerodynamic derivatives of a flexible aircraft using output error method, in *IEEE International Conference on Mechanical and Aerospace Engineering (cmae-2011)* (2011), pp. 361–365

R.K. Mehra, Maximum likelihood identification of aircraft parameters, in *The Joint Automatic Control Conference* (1970), pp. 442–444